LIFE IN THE UNIVERSE

OTHER WRITINGS
BY MARSHALL VIAN SUMMERS

THE GREAT WAVES OF CHANGE

GREATER COMMUNITY SPIRITUALITY

STEPS TO KNOWLEDGE

RELATIONSHIPS & HIGHER PURPOSE

LIVING THE WAY OF KNOWLEDGE

WISDOM FROM THE GREATER COMMUNITY
Volumes I & II

THE ALLIES OF HUMANITY Books I-III

THE REALITY AND SPIRITUALITY OF LIFE IN THE UNIVERSE

Marshall Vian Summers

NEW KNOWLEDGE LIBRARY

LIFE IN THE UNIVERSE

Edited by Darlene Mitchell
Book Cover Photo Image, "My God, it's full of stars!"
by Thomas Zimmer (www.thomas-zimmer-photography.com)

ISBN: 978-1-884238-49-9
Library of Congress Control Number: 2011945655

Publisher's Cataloging-in-Publication
(Provided by Quality Books, Inc.)

Summers, Marshall.
Life in the universe / by Marshall Vian Summers.
p. cm.
LCCN 2011945655
ISBN 9781884238499 (pbk.)
ISBN 9781884238994 (hardcover)
ISBN 9781884238406

1. Cosmology. 2. Spiritual life.
3. Extraterrestrial beings. I. Title.

QB982.S86 2012 523.1
QBI12-600050

The books of New Knowledge Library are published by The Society for The New Message. The Society is a religious non-profit organization dedicated to presenting and teaching a New Message for humanity.

The books of the New Knowledge Library, of which *LIFE IN THE UNIVERSE* is one part, are being studied by people around the world in over 15 languages and can be ordered at newknowledgelibrary.org, amazon.com and at your local bookstore.

To learn more about The Society's audio recordings, educational programs and contemplative services please visit newmessage.org or call 1-800-938-3891.

The Society for The New Message
P.O. Box 1724 • Boulder, CO 80306-1724 • (303) 938-8401
email: society@greatercommunity.org
internet: newmessage.org

DEDICATION

To all those who feel called
to go beyond the limits
of human existence.

TABLE OF CONTENTS

PART TWO
THE SPIRITUALITY OF INTELLIGENT LIFE IN THE UNIVERSE

Introduction

How often does a book arrive on Earth that could change the course of human history? As the Bible and the Quran have done, what new revelation could change how we see ourselves, our world and our destiny?

What you hold in your hands is much more than a book. Life in the Universe is a revelation. Across 173 pages and 4,749 lines of text, this revelation details the interactions of civilizations in our region of space, the challenge of facing a non-human universe and the spiritual dimensions of all intelligent life that has evolved since the beginning of time.

How could any writer, thinker, scientist or philosopher claim to have any direct knowledge of these things? Setting aside speculation, the instruments of science and the pen-pad of science fiction, what do we really know about life in the universe?

In truth, very little. Like tribesmen looking out to the dim horizon beyond an isolated oceanic island, we know nothing about life beyond our shores. We look up to the stars, peering into the deep black of space, wondering who is out there. We simply do not know. And as this book will reveal, our unknowing is one of the great perils of our modern time.

We build arrays of radio telescopes, hoping to hear messages from space. We hone our optical instruments, hoping to discover Earthlike worlds. But what if we are the ones to be discovered? What if contact with extraterrestrial life happens on our shores? And what if it happens on their terms, not on ours? History gives us a dire warning. It is the unfortunate position to be the race that is discovered, to be the natives of a new world.

As some in the corridors of government and commerce are well aware, we have been discovered. Our encounter with extraterrestrial life is well underway. Contact has begun, but it is with a dark and self-serving intent, occurring without human consent or awareness.

This is why you hold in your hands a book called LIFE IN THE UNIVERSE. This book is God's response to the great vulnerability of human civilization as it stands at the threshold of space. Can the trickle-rate of scientific progress prepare us for our present encounter with life in the universe? For this, a new revelation is needed.

LIFE IN THE UNIVERSE is part of a vast education and preparation for humanity called the "New Message." Over 9000 pages in length, the New Message is a Divine answer to the panoply of global problems facing humanity: destruction of our natural environment, depletion of Earth's energy and life-sustaining resources, escalating religious and political conflict and intervention by certain races in our region of space.

If God speaks to the world at those times when humanity's needs are most pressing, then this might be the most justified moment in history for God to speak again.

LIFE IN THE UNIVERSE is the exact rendering of a revelation event. The 173 pages of this revelation were received over a three-day period in 2008 by a man named Marshall Vian Summers. After leaving his work as a teacher for the blind, Marshall embarked on a mysterious path of preparation that lasted over a quarter century. Marshall has spent over 30 years in relative seclusion preparing to receive a New Message for humanity, of which LIFE IN THE UNIVERSE is a part.

The revelation event began the morning of May 11th. And

by the afternoon of May 13th it was over. Eleven hours of speech, separated by two nights, had etched onto digital tape the words of a revelation destined for the world. Within a few days, the original recordings were transcribed and then printed on simple paper. In these mere 11 hours, without revision, editing or the authoring of a single word, a book had arrived on Earth with the power to change the course of human awareness and understanding.

The reader should be prepared for the experience of reading this book. Its message is uncompromisingly direct. It says what is, not what might be. And it challenges the many assumptions and fantasies about life in the universe that are commonplace and popular in human culture. What we might want out of this book, or out of the universe as a whole, seems left out entirely or flatly refuted.

At times reading this book, you may find yourself experiencing something like "astro-depression." How can the universe be this worldly? Where are the enlightened races, the stories of great empires, heroes, free energy and adventures in space? Why can't it be better than life here on Earth?

It is natural for us to want these things, even to demand them from the last remaining frontier of existence. For all of human history, the universe remained comfortably unknown, allowing our fantasies and mythologies to grow and be untempered by reality. The universe became the canvas upon which we expressed our longings and our dissatisfaction with life here on Earth. Given our isolation in the universe, this is understandable.

Yet now it is time to face the living universe. And doing this is a courageous act. Here we must be willing to venture beyond the comfort of our beliefs and assumptions and bravely look into a universe that we know little about. This is the great evolutionary step before us.

The words of LIFE IN THE UNIVERSE are simple and unadorned. Its message can be blunt and to the point. This is especially true in Part 1: The Reality. We might find some of its phrases awkward and wonder to ourselves why God doesn't speak more majestically, more perfectly, more spiritually. And yet perhaps this is what revelation really looks like. The Biblical and Quranic texts we associate with revelation have passed through time, translation and the innumerable hands of followers, scholars and rulers. Do we know what original revelation is really like at the time it is given?

Furthermore, we should keep in mind that LIFE IN THE UNIVERSE was not written to be read. It was spoken to be heard. It is a spoken document and has not been adjusted to the literary tastes of anyone. It is pure, untainted by any human hand or intellect, set down on paper exactly as the recorder captured it at the moment of revelation.

While LIFE IN THE UNIVERSE may disappoint some of our expectations and open rain clouds upon the parade of human preferences and assumptions, it is nonetheless a beautiful revelation of breathtaking scope. It opens up a vast and panoramic vision of life on the grandest scale. It is truly a doorway into the living story of our universe.

It is in the lineage of great Messengers that my father has brought to us something never before seen on Earth. Yet it is not only a social, moral or metaphysical need that has compelled God to speak again. LIFE IN THE UNIVERSE has come to meet a greater human need: to restore our world, to unite the tribes of Earth and to prepare for our emergence into the universe, the vast arena of life which encompasses the other 99.999...percent of Creation.

It is in this new frontier of life that we must build our future. It is here that our success and survival will be determined. This is the greatest threshold we have ever faced. But we are not prepared.

The education and preparation have been given. The doorway to life in the universe stands before us. No longer do we look through the keyhole, imagining what lies beyond. In our hands, we now have the key itself: a simple book, authored by no human on Earth, with the power to prepare humanity for our future and destiny in the universe.

Join me in using this key to open the door to a new chapter in the human experience. Let us pass through together, leaving our assumptions and expectations at the door, and prepare for life amongst the Greater Community of worlds.

Reed Summers
Boulder, Colorado 2012

PART ONE

THE REALITY OF INTELLIGENT LIFE IN THE UNIVERSE

Within the
Greater Community,
you are but a beginner.

CHAPTER 1

Facing a Greater Community of Life

You live within a Greater Community of intelligent life in the universe. It is vast, encompassing the entire spectrum of evolution—the evolution of technology, the evolution of societies and the evolution of spiritual awareness and ethics. It is vast and incomprehensible. It is the greater environment into which humanity is now emerging. Humanity is emerging into this environment not from its brief forays into your local solar system, but from visitations that are occurring in the world now and which will increase in the future.

This greater environment into which you are emerging is a physical reality governed by physical laws, the laws of nature that you are aware of. Do not concern yourselves here with the possibilities of other dimensions, for that will not be your need and concern. It is this greater reality in which humanity has always lived and in which your world has always existed that must become your focal point now.

Many people have speculated on the possibilities for life in the Greater Community, what that might be like and how life would evolve into more sophisticated realms and societies. Today, as in the past, people project their hopes and their wishes upon the meaning of life beyond the world, hoping that other races will be more enlightened, more sophisticated and more elevated than the human family has been and is currently. And, of course, there are many fears projected into this larger arena of life—fears of terrible beasts, terrible invading nations, destruction, attack and so forth.

But, as always, reality is very different from expectation. The reality into which you will be emerging will be the focus of this series of teachings—a reality that you can understand from your own experience of living in the natural world and from your own experience of living within the evolution of human society.

The great difference, of course, is both the complexity and the vastness of this Greater Community and the fact that it is inhabited by races of beings that are very different from humanity, not only different in form and appearance, but different in intelligence and awareness and different in ethics and values.

This will be hard to deal with, and this is where your hopes and fears may arise. But you must recognize that humanity has now reached a point of world development and a threshold where it will be facing Great Waves of environmental change and difficulty in the world. Here your encounter with intelligent life from the Greater Community and the reality of the Greater Community itself will become ever more important and central to your well-being and to the kind of future you will be able to create.

There is a very great shift here, a very great shift from a human-centered world to a Greater Community reality. Because you have lived in isolation and have evolved in isolation for so very long with only very infrequent and largely unrecognized visitations to the world from various races in the past, you naturally project your values onto the prospect and the notion of intelligent life in the universe. As a result, some people feel or believe fervently that the universe is inhabited by human beings who have evolved beyond conflict and deception and that the values that you hold here or that you aspire to are universal in nature. The fact that this is not true and that it represents a dangerous set of assumptions and beliefs is something that you must face at the outset.

The other difficulty is that you are not preeminent within this Greater Community though you have established preeminence within your own world. Within the Greater Community, you are but a beginner. You are standing at the beginning of your development, unaware and possessing assumptions and beliefs that will not be true. You have great disadvantages and are vulnerable to persuasion and deception from various races, both in the future and from those who are indeed intervening in your world at this time. Your isolation has given you a naiveté and has not prepared you for the challenges and the opportunities that the Greater Community itself will present.

There is no one in the world who could teach you about the reality and the spirituality of life in the universe. There is no one who has this awareness or who has this experience. As a result, the communication and education about this must come from God and, secondarily, from those very few races who are aware of your existence, who will support your emergence as a free and self-determined race and who will be called by God to assist you in this most fundamental and necessary education.

You have allies in the universe though they are not visiting your world, for humanity is not ready to establish relations with other nations in the universe. Humanity does not have the maturity and is not united, strong enough or discreet enough to engage meaningfully and beneficially with any race in the universe.

This is not a time where Contact is required. It is a time where preparation for Contact is required. And this Contact will take a great deal of time. It will take time because an education of this magnitude will not come quickly. Many people will resist it, deny it or avoid it. This education will not come quickly because humanity must now concern itself ever increasingly with the Great Waves of change that are coming to the world and that are here already—environmental degradation, the depletion of your fundamental resources, the loss of food production, the scarcity of water, the dramatic effects of a changing climate and a heating world and the ever-growing risk of competition, conflict and war between groups and nations over access to the remaining resources.

This will become a dominating focus and concern, both for average people in the world everywhere and for nations and governments as well. However, to meet these immense and unprecedented challenges, you will need Knowledge and Wisdom from beyond the world. Here it is not technology you need as much as it is the will and the commitment to unite for the preservation of this world as a habitable environment for the human family and to cease your endless conflicts to prepare to engage with intelligent races in the universe, many of whom will not be here to support you.

Meeting the Great Waves of change, establishing the necessary human cooperation and the cessation of human conflict along with the preparation for the Greater Community represent the great and fundamental needs of humanity. If these great needs are not sufficiently recognized and addressed, then everything else that you will attempt to create for your own benefit, either personally or for the benefit of humanity, will prove to be insufficient, and great travail will come to the world.

Your education about the Greater Community must come from God. It must come from God to be pure and to be wholly beneficial for you. And it must come from God because only God knows fully the nature, the reality and the purpose of the human family. No foreign power or race could really know this, even if they studied your behavior and your transmissions. The communication must come from God, and that is exactly what is happening.

The teachings that you are about to read concerning the reality and the spirituality of the Greater Community come from God, for there is no one in the world who could know the things that will be presented here. And there is no foreign power in the universe who could communicate what humanity really needs with thoroughness and complete knowledge of human nature, purpose and reality.

Accept then that God has sent a New Message into the world to prepare humanity, both to face the Great Waves of change that are coming to the world and to prepare for the reality and spirituality of life in the universe and all of its difficulties and hidden opportunities. This is an education that you cannot give yourself. It is an education that you could not receive in any university or center of learning in the world. It is a communication that you must listen to with your heart as well as your mind.

Your mind will be confused by many things that are presented here, and many of your ideas will be challenged and shown to be incorrect or inadequate. Your mind may reject this great teaching and preparation. It may object for reasons both rational and irrational because your mind does not know the mind of God. Your mind only has fixed notions about life in the universe. Yet it has not experienced life in the universe.

That is why you must receive this teaching and education about the reality and the spirituality of the Greater Community with both your mind and your heart. Your mind will struggle with the ideas and perspective presented here, but your heart will know. You will know because it is God speaking to you, educating you and preparing you. Through this education and preparation, God is giving you greater strength, greater security and a greater protection that you will need in order to face the challenges within your own world and the challenges that surely exist beyond it.

Many people claim to have knowledge of the universe, but how can they know? They have been trapped on the surface of this one world—believing, hoping and speculating; trying to comprehend past revelations and prophecies or trying to project their understanding into realms far beyond their experience or awareness. Complex theories may be established. Prophecies from the past may be attempted to be fulfilled in current times. But these must all prove inadequate, for humanity is without education regarding the larger universe in which you exist and which you will have to face increasingly, both now and in the times to come.

It is wise then to accept your limitations. It is wise to put yourself in a position to be a student, to be a listener and a learner so that you may receive fully the revelations that will be presented here—the revelations born of a New Message for humanity, revelations that will provide for humanity a new way forward into a very difficult and uncertain future.

Those who will receive and accept this challenge of learning about Knowledge and Wisdom in the Greater Community will be the great beneficiaries, not only for themselves personally, but for the entire human family. They will be the ones who will be in a position to educate and to prepare others. This will require vision, courage and great objectivity, for this education must speak through and beyond your hopes and your fears. For they can only cloud the vision that you must now have.

Accept that you do not know what life in the universe is like, how it exists, how it interacts with itself, what to expect regarding visitation to your world and what to understand about the visitation that has already occurred here. Accepting this limitation gives you the greatest opportunity and possibility to see beyond your current limits, to see beyond the confines of this world and to see beyond the confines of human interactions. For now you will be considering and facing interactions of a very different nature with intelligences that are very different, who think very differently and function from very different assumptions about life than you do or that humanity as a whole does.

This is a preparation that will make all the difference in determining whether humanity's initial engagements with life in the universe will be beneficial or will be detrimental. Much wisdom must be brought to bear here, for you are the stewards of a beautiful planet that is greatly valued by others. But you are weak stewards, and you are conflicted amongst yourselves. And you are rapidly depleting the wealth of this world, a world that is so rare and valuable in a universe of barren planets.

You do not yet see your vulnerability here nor the importance of maintaining this world, maintaining your self-sufficiency, maintaining the world's climate and maintaining its natural balances. Do not think that you can travel to other worlds and find places like this. They are very rare, and they are always occupied. This understanding will give you a greater sense of responsibility, for real education must always imbue you with a greater sense of responsibility—responsibility not only to learn, but to act and to serve.

Humanity now is on a reckless course. It is destroying its self-sufficiency in the world. It is endangering the natural balances that have given humanity such a beautiful and splendid world in which to evolve and to live. This is hazardous not only for your future in this world, but also hazardous regarding your relationships with other races in the universe.

Through this series of teachings that you are about to experience, you will be able to learn about this, to see its implications and to understand this reality in order to have a real sense of it. This recognition of life in the universe is possible because you were born with a Greater Community awareness. This awareness gives you the ability to empathize and recognize certain realities about life from beyond this world. This is a recognition that exists far beyond your intellect, your thinking mind, which has been cultivated and established through cultural conditioning and your worldly education.

Recognizing that you have a connection to life in the universe is very important here and will be part of the exploration within these teachings. Without this deeper connection, the universe would be too awesome, too immense, too complicated, too difficult and too threatening. It would produce only insecurity and confusion for you. But speaking to the deeper part of you enables you to gain recognition and a sense of your destiny within the Greater Community itself. For regardless of the difficulties that you face in your own world and the difficulties in establishing proper and wise engagements with races from beyond your world, there is the fundamental reality that the human family has a destiny in space. You have a destiny to engage with other races, a destiny to grow and, with the proper education and understanding and with your own native wisdom, the possibility of being a free and self-determined race in a vast and difficult universe.

Accept this challenge, recognizing as you do that there is something within you that gives you the ability to respond to the revelations that will be given here. This ability exists beyond your intellect, beyond your ideas, beyond your fixed notions and beyond anything that you have yet learned in the world. It is mysterious because it is born of God. It is born of a Greater Reality of which you are a part.

Not everyone will be able to see this and know this. And it is important even at the outset that you accept that many people, even those you are very close to, may not be able to see, to know and to respond to this preparation for the Greater Community. But that is all right. It is given now for the first to respond. If you are amongst the first to respond and amongst the first to be genuinely educated about life in the universe, then you must accept this challenge—even if others do not respond, even if others are fearful or doubtful, even if others turn away.

Not everyone is at the same point in their development, education and awareness. Therefore, do not think of everyone responding. Instead, think of you responding and learning. Higher education is not for everyone at this moment, and this most certainly represents a higher education.

Therefore, prepare yourself to receive. And if you hear things that you do not understand, then be patient, for in the course of these revelations We will repeat things many times and describe things in different ways to give you the greatest opportunity to see, to know and to recognize the immense environment of life that exists beyond your borders and of which you have always been a part.

You are the stewards
of a beautiful planet
that is greatly valued
by others.

CHAPTER 2

A Brief History of Visitation to the Earth

To understand the present and to be prepared for the future, there are certain things from the past that must be clarified and that you must learn to see. The first is that the world has been visited for a very long time. These visitations have been from a number of different races for different purposes. Some have come here to gain biological samples from the immensely rich diversity of life in the world. Others have come here to hide things, to store things beyond the awareness of the native peoples of the world. At different points, others have established bases here temporarily and on a few occasions have tried, in more recent history, to educate or to influence early human civilizations.

These visitations were brief and were not sustained for long. The bases that were established here were only mainly to gain a greater understanding of the Earth's geological and biological realities. Attempts to influence early civilizations proved to be unsuccessful. However, a great deal has been learned about the natural world here, and many biological elements have been taken from the world periodically to support the emergence of life elsewhere.

Those races who are present in the world today represent a different set of forces from those who have visited humanity in the past. Humanity has lived in isolation throughout its entire evolution with only very brief encounters with races from beyond the world. Primitive peoples had a great capacity for recognizing the forces of nature, but their sophistication and understanding, technology and communication were limited.

Therefore, with only a few exceptions, attempts to communicate with them were never made. They were deemed unready to engage in meaningful dialogue. Yet certain native groups were given wisdom about life existing beyond the world. This was demonstrated to them by visitations and by the giving of gifts. This existed in tribal nations long before and, in some cases, after the existence of early human civilizations.

Since technology is a very important part of life in the universe, particularly regarding travel and communication, attempts to influence these early civilizations proved to be unsuccessful. It would have to wait until a much later time for humanity to develop technologically and to develop social structures sophisticated enough and sufficiently large enough for greater communication to be established.

For nearly all the races who visited the world in the past, and for those races who are here at present, the world presents a very great challenge. Living in space for very long periods of time in sterile environments, races coming to the world are faced with an immense difficulty. The degree and the diversity of biological agents in the world make this planet, though extremely conducive to life, very hazardous for any visitor who is not adapted to it. Even races with very advanced technology cannot protect themselves against the diversity of life and the number of biological agents that exist in this terrestrial environment. This has made actual living in the world extremely difficult and, in most cases, impossible for all the races who have visited here. It would take them generations to adapt to living in an environment such as this. And only through a prolonged genetic program of bonding with the native people could they have a real possibility of living in a world of such diverse and complex biological forces.

It may seem strange to people that the world would be so hazardous for an advanced race who did not evolve here. But it is true nonetheless. If you live in a sterile environment, you can visit other sterile environments, or you can visit environments where there is a very limited diversity of life, if you are prepared for them. Biological creatures are vulnerable to contamination everywhere in the universe. There is no technology, no medical technology, that can protect a biological entity of any kind from new and unexpected influences. This has made your world an extremely difficult place to visit for any length of time or to live in without a very advanced technology. And to maintain this technology far from the source of one's own civilization is very difficult.

The world does not provide the technology and the technological inputs that would be necessary for a foreign race to sustain a technological environment here for very long. And the hazards of contamination would be extremely great. For no matter how advanced one's technology, one can still succumb to unseen biological agents such as bacteria and viruses. This makes the approach to a world such as yours a very difficult proposition for any race. Even other races who live in environments of biological diversity, and who have been able to maintain those environments and not strip them of resources, would find this to be very challenging.

Visitations to the world have been brief and infrequent, carried out by different groups for different purposes. Some have come to gain biological resources, some for experimentation and some to establish a base of influence. But none of them have been able to be sustained for very long.

As a hiding place, the world has offered certain races great advantages. Being densely covered with foliage in many places and inhabited by native peoples who were superstitious and whose existence was very local within certain areas has made it possible to hide things of great value here, and certain races have taken advantage of this. Even today there are things hidden in the world beyond human awareness that have been here for a very long time.

However, now as humanity's technology is rapidly accelerating, the risk of these things being discovered has increased. Many of these treasures buried here for ages have been removed from the world in recent times, especially in the last century. Humanity's ability to discern the existence of objects—metals and machinery deeply buried even in the ocean—is increasing rapidly, and this has required certain races to return to the Earth to remove things of value that have been stored here for a long time.

Despite the difficulties of visiting or dwelling in this world for an advanced technological race, this world has been the focus for many races for a very long time because of its biological diversity and its strategic position in this well-inhabited part of the universe. The Earth has been observed for a very long time. It is of great interest to different groups for different purposes.

The ascension of humankind and the rapid development of human technology over the past two centuries have brought different forces to the world, forces who are looking for the possibility to gain advantage and to unite with humanity to gain control of an emerging world such as this. With few exceptions, this was never the focus before, for humanity lived in a primitive state and had not yet built any technology or system of communication that other races could use or benefit from. Native peoples were only of interest in terms of their adaptive capabilities and their knowledge of the local environment. Beyond this, there was no possibility of really uniting with them. The attempt at uniting with human beings genetically, therefore, has only been a very recent phenomenon.

In the past, the Earth had been viewed as a storehouse of biological wealth. Yet attempts to establish permanent residences here have failed as the visitors became affected by the biological environment. Even their advanced technology eventually could not protect them from these forces of nature. And the possibility of introducing foreign biological agents into their planets of origin and into their bases in space proved to be far too great, grievous and risky for settlements to be established here for very long. For should a biological agent infect members of an expeditionary force, they could carry these agents to their centers of operation, thus affecting everyone involved.

People have adapted to the world's environment to a very large degree, but in the history of the movement of peoples within the world, you have seen the power and the impact of infectious diseases in destroying entire populations as people from different continents came in contact with one another.

This is a very great hazard in space and continues to be a great hazard. Technological races function in largely sterile environments. They engage with each other in sterile environments. Commerce between nations is rarely carried out on the surface of their worlds unless those worlds have been stripped of their biological agents, or they are planets that never had the presence of a biological evolution. They are simply being used as bases of operation. In the affairs of engagement and commerce in the universe, contact is nearly always made on board craft of some kind or in environments that are well protected and secure from the invasion of other biological agents.

The spreading of viruses has been a serious problem in the past in many regions of space, particularly in well-inhabited regions such as your own. This is one of the reasons why humanity's future ventures into space are viewed with such trepidation. The possibility of engaging with human beings and being exposed to such an unprecedented and unexpected degree of biological influence is looked upon with great concern. This is why anyone visiting this world, even currently, cannot live here and must take great precautions to protect themselves against the biological environment of the world.

In the affairs between nations in the universe, contamination is a very serious problem, particularly concerning races who have evolved in very different environments and who carry within themselves biological agents that could be extremely detrimental to others. As you will see through the course of these teachings, technology does not erase all the hazards of life and, in many cases, has increased those hazards significantly. For races who live and travel in sterile environments, this risk is immense. Contamination is a major concern for all races who travel in space and who engage in commerce with others.

Therefore, while the world has been greatly valued and visited many times, the attempt to live here and the attempt to genetically bond with races who live here has proven to be extremely difficult and has not been attempted for very long. Human folklore and human mythology can reveal any kind of image, but the reality remains that the Earth has been looked at as an extremely hazardous, though rich, environment by those very few races who are even aware of its existence.

Another reason the Earth has not been visited very often or by very many different groups is the problem of travel in space itself. There are people today who speculate that through inter-dimensional travel you can go anywhere in the universe you want. But in the experience of advanced races in the universe, such attempts have proven to be extremely dangerous and unfortunate. Getting around is a lot slower than you might think. Most races that travel in space only do so within local regions. They only travel in local regions because it is extremely dangerous to enter into a territory governed or overseen by others. And if you venture too far from your home planet, you would not be able to sustain yourself over the course of time.

There are many districts where travel and trade are restricted and are off limits to others. Therefore, you cannot travel freely in the universe unless you are in a region where there is a very sparse or limited development of intelligent life and where there are no restrictions to travel and trade.

In the region of space in which your world exists, which is highly inhabited compared to other regions, there are great restrictions as to where one can travel, where one can visit, whom one can contact and for what purpose. You cannot go visit any planet you want, for that is a violation of others' territories and those regions where others have specific interests. You cannot travel along primary trade routes without permission from governing bodies.

This is a very complex situation about which humanity knows nothing at all. People think the universe is just a huge empty place awaiting exploration, containing new worlds with vast resources—there for the taking. But you would not have to travel far beyond this solar system to discover that other places of value are already owned by others and that long-standing rules of engagement have been established as to who can visit these places and who has priority over them. And because, in your local universe, nations have created such long-standing agreements and have established stability between nations and worlds over a long period of time, these rules of engagement are very fixed, though they remain unknown to you who have yet to travel beyond your borders.

Therefore, not anyone from anywhere in the universe can come to the Earth because the Earth exists in a well-inhabited region that is governed by others, where trade and travel are restricted and overseen by governing bodies. Even within this district, a district being a region of space, there are very few races that are aware of the existence of this world, for they never travel in this direction. Those who view the world with ambition will not tell others of its existence, seeking not to have further competition regarding the future of this world. For many races who are aware of your world, it is a secret—a secret that they keep to themselves, not wanting others to know about the existence of this beautiful, biologically diverse planet, which is ruled by a weak and conflicted set of tribes and groups.

As a result, the number of races in the universe who are aware of this world has remained very limited. If your world existed in a highly uninhabited part of the universe, anyone could come here for any purpose and do whatever they wanted. But that is not the case with your world. Other races from beyond this district cannot travel in these regions, and the secret of this world is kept and guarded by those who are aware of your existence and who have designs for your future.

Humanity's engagement with the Greater Community, therefore, has been extremely limited to the point of almost complete isolation. Those races who have visited the world have not sought to reveal to the native peoples or even to the modern peoples of the world their purpose, their technology or their intentions. Those races who are here to take advantage of a weak and divided humanity certainly would not do that. For you to expect or demand that they do so represents a naiveté and a lack of education and awareness on your part. Even your potential allies in the universe, who value the possibilities and the potential of the human family, even they do not want to let other races know of your existence for fear that there will be even more intruders here, leading to an even greater danger to the future and the freedom of an emerging humanity.

Humanity remains uncontaminated. It remains an isolated race. It is evolving on its own, and this evolution has been accelerating. But the acceleration of this evolution is not because of the infusion of extraterrestrial technology, but because of humanity's own trajectory in nature. Only a few technological items have been introduced to the world surreptitiously to advance humanity's communication systems by those races who seek to use these in the future for their own purposes. But, by and large, humanity's evolution and technological advancement have been very slow up until quite recently and have primarily been the result of human ingenuity and invention.

However, humanity remains a very primitive race with primitive tendencies and tribal animosities. Other races in the universe recognize this and fear such tendencies would influence their own nations, the majority of which are hierarchical and united, where personal freedom is unknown and greatly feared as an undermining force.

Contrary to some people's expectations that you will meet highly advanced, free and peaceful nations, most societies that are aware of your existence and that have an interest in this world function under a strict hierarchy that you would find intolerable. Only your few allies that exist in this region represent independent races, and their independence has had to be guarded very carefully. For freedom is rare in the universe, as it is in your own world. This is the great truth that you will have to face, in contrast perhaps to your expectations, hopes and wishes.

This accounts again for the relative infrequency of foreign visitation to the world. Those visiting here do not want the influences of tribal cultures to affect their own social structures. Even the existence of music and dance, which have been so much a part of culture around the world, is looked at by some nations with great anxiety and fear, not wanting such influences within their own social structures. For in visiting a world, you will influence it, and it will influence you. Visiting races will influence you, and you will influence them. And here, influence, even by very primitive tribes of people, has been looked at with some degree of anxiety.

It must be understood here that the value of humanity is only regarded by a very few races in your local district of space. And those races who do value humanity for its own sake—for its accomplishments and for its potential—represent a small minority of the few races who are aware of your existence. Those that do could be considered allies of humanity insofar as they would support and try to protect the natural emergence of humanity against the influence of foreign powers who would corrupt humanity and who would seek to bring humanity under foreign domination and control. This is the dilemma of living in a world of such immense value. Humanity has now gained a sufficient technology that other races can use, and so the vulnerability of humanity to foreign persuasion and intervention now is very great.

Your past experiences with the Greater Community cannot really help you. They were extremely infrequent, and the only record of them now has been passed on through oral tradition or, in rare cases, writing. This history has become so modified by human ritual and belief that it holds no accurate or direct record of intervention or foreign presence in the world.

Therefore, you cannot look to the past to understand the nature, the purpose or the reality of the foreign presence and intervention in your world today. You have only a few traces of evidence to tell you that visitation has occurred, largely through the writings and artwork of native peoples from different cultures, from different eras. But this is insufficient as a guide. This will not reveal to you the reality of the universe around you, the forces that exist there or what humanity will have to face and deal with in the future.

Look at the stars and space
not as a refuge,
but as a representation
of life on a larger scale.

CHAPTER 3

The Limits of Space Travel

As has been previously indicated, travel in space is relatively slow. And inter-dimensional travel has proven to be disastrous for most races who have attempted it. Though travel in space is extremely rapid by your standards and by the degree of your own scientific development, getting around is still very difficult and time consuming. Because travel and trade are so restricted in highly developed parts of the universe where there is a great concentration of advanced nations, movement is restricted and restrained. You cannot go anywhere you want under these kinds of circumstances without violating others' territories or without violating the rules of travel and trade.

Going from one end of the galaxy to the other is just simply out of the question. It is a fantasy to think that this would be the case. Traveling through dimensions has proven so hazardous that there are very few races in the universe who even attempt it at this point, for those who enter these thresholds never return and are never accounted for, and there is no way to go find them and learn what happened to them. The dimensionality of space is so complete and the boundaries between dimensions are so significant that to explore these kinds of things has proven to be overwhelmingly difficult. Even where it has succeeded, races have emerged in other races' territories or have entered hostile physical environments and have never been able to find their way back.

The galaxy as a whole is uncharted. Only regions within it are charted. Only regions where there is a great deal of trade, travel and the existence of evolved technological nations are charted. Beyond this, you have uncharted territories, sometimes immensely huge, where any traveler could become lost without known reference points.

Ships that travel use fuel. Fuel has limits. Even fuels that are derived from solar energy have limits. Even fuel that is derived from nuclear power has limits. If you go too far, you cannot return. If your exploration takes you too far afield, you cannot return. If you enter an uncharted region, you will face physiological hazards and the possibility of entering another's territory who could prove to be hostile to your presence. You certainly could become lost in uncharted territories, as many travelers have.

To extend one nation's reach is also very difficult because any establishment that you make far away must be reinforced. Even if you position yourself in an environment that is relatively inhabitable from both a physiological and a biological standpoint, reinforcing foreign establishments is very difficult and requires a tremendous amount of travel and resources. Distant mining colonies are very vulnerable to piracy. They are very vulnerable to breakdown. They must be sustained with a great deal of effort even if the resources they are discovering prove to be of immense value. As a result, most traveling nations stay fairly close to home and rely upon trade to gain distant resources and rely upon local commerce to gain what they need.

Of course, human beings with their wonderful imagination conjure up all kinds of wonderful ways to travel around the universe almost effortlessly, using energy that is somehow derived from physics and can sustain unlimited travel, even collapsing time and space. But for those who live in the real universe, they have had to face the limits of technology. In some cases, these limits are quite severe.

Races who travel generally create their own food sources, but even this requires the input of resources. Everyone in the physiological realm must still eat and take in energy in one form or another. They must use energy and gain the resources for producing this energy. They must rely upon technology, which must be reinforced and supported. If they travel very far from their home planet or their home base of operations, they face a tremendous number of logistical difficulties.

In areas of space where there are not high concentrations of advanced nations, you have tremendous difficulties with piracy. You have tremendous difficulties with becoming lost or becoming subjected to hostile local influences or responses to your presence. There are regions in space that are simply considered too dangerous to travel in because of these reasons. Even areas that are sparsely inhabited, where routes and coordinates have been established for travel, even here there are extreme hazards. Over the course of time, nations and even small empires establish a stability that they maintain and sustain on their own base of resources and influence.

There is no great empire that rules the galaxy. That is a complete fantasy. There are very strong local powers and associations of power, often established through trade unions or governing bodies that oversee trade routes. But immense empires that govern vast regions simply do not work, for there is no way to maintain control over the course of time, and there are too many challenges to disrupt a structure of this size. Yet there are empires that contain dozens of planets, star systems and so forth. This is considered fairly large, particularly in your local environment.

In the universe, if you extend your influence, you must control that influence and you must sustain that influence. You must sustain it through technology. You must sustain it through resources. And you must sustain it through constant effort to maintain foreign establishments. Beyond a certain point, this proves to be too difficult, too draining of a nation or even a collection of nations to support and to sustain.

As a result, there are large regions of the galaxy that have been unexplored and rarely ever visited by anyone. Though there are races existing within these regions, they tend to be non-technological. If they are technological, then they tend to be extremely limited in their range. Unable to trade or gain advanced technology from other nations through the avenues of trade and commerce, they remain extremely isolated.

This is the universe that you live in—a universe of magnificent creations, but also a universe of tremendous restraints. Resource acquisition is a problem for all advancing nations. Technology requires resources. The greater the technology, the greater the resources that are required. The larger one's sphere of influence, empire or ownership of property, the greater the requirement for resources.

Nations that are forced to depend upon trade lose much of their self-determination in the process. Unable now to provide for themselves adequately, they must rely upon foreign provisions as well as foreign political influence to sustain them. For a nation that seeks to be free and self-sufficient, the problem of trade and dependence upon others is extremely great. That is why the three fundamental requirements to be free and self-sufficient in the universe are unity, a united population, self-sufficiency and extreme discretion.

If you have wealth in the universe, others will want it. They will either want to take it from you or they will want to trade for it. If they trade for it, they will try to seduce you or convince you that what they have to offer is something you really need, even if you do not really need it. The problem with having wealth—whether it be biological wealth, technology that others do not have, ownership of a strategic world or, as is the case with Earth, a world of tremendous biological diversity—is that maintaining control is difficult if others are aware of you. This becomes impossible if you depend upon others for your fundamental resources.

Where there is trade, there is influence. The extension of influence in the Greater Community has reached very great degrees of sophistication and potency. For a young emerging race such as humanity, the Greater Community represents a very hazardous environment. The wealth of your world, from which you have benefited for so long and from which you benefit today, gives you a greater vulnerability to foreign interest and influence. And because humanity is so easily influenced, by Greater Community standards, this puts you in tremendous jeopardy.

Therefore, your expectations of trade and travel must be constrained by the realities themselves. You will be able to travel freely within this solar system as long as you do not try to discover anyone's foreign establishments here. But if you move beyond this solar system, you are entering a larger arena of intelligent life where territories and routes of travel have been long established. You will not be able to go wherever you want, to visit wherever you want or to take whatever you want. The idea that humanity will go out and search an empty universe for resources is completely false and really represents a very dangerous and even fatal assumption.

As long as you are weak and powerless, others will not try to defend themselves against you. But if you become aggressive and ambitious, then you will run into tremendous problems. For in the environment of space in which your world exists, there are long-standing contracts and arrangements that you will not be able to change. Should you violate them, you will have to face a whole array of opponents far beyond your skill and ability to contend with.

That is why maintaining self-sufficiency on Earth is so fundamentally necessary to preserve humanity's freedom and self-determination. If you can thwart inappropriate interventions in the world, provide for yourself adequately, not be overly greedy or ambitious and be satisfied with the splendor of what the world provides you, then you can have great immunity here, and others will respect this. But if you continue to violate the world—destroying its fundamental resources, using up its wealth, creating greater instability and conflict and driving the life-sustaining biological systems into instability—then others will feel compelled to intervene here to save this world for themselves. Indeed, this is what is happening today.

Do not think then, should you deplete the world, that you will be able to go out and get whatever you want from other splendid worlds such as this, for they will be inhabited, controlled or overseen by others. And you are in no position to mount a conflict with other nations in your region.

In well-established and well-inhabited regions of space, war is very rare. Internal conflicts certainly do erupt, and there are changes in leadership and administration in various worlds, sometimes even through violent means. But conflict between worlds that trade and interact with each other is very rare.

Over long periods of time, war is seen as mutually destructive and literally is prevented from occurring between nations, particularly nations that have important resources and who are powerful within their regions. This tempers ambitions and the desire for conquest, for such ambitions prove to be a detriment to everyone involved. Should you prove to be ambitious, desiring conquest, others will amass against you to such a degree that you can never oppose them.

Stability is more important here than the advancement of one's empire—stability and the steady flow of resources—a kind of status quo, you might say, that has proven to be beneficial over a long period of time. For a young, aggressive race such as humanity, it is hard to understand this. Humanity is like a wild adolescent who has great passions, great ambitions and great plans, but is reckless and self-destructive. Humanity has not yet emerged into a mature environment where such passions and recklessness are not tolerated.

You live in a universe full of restraints. It is not a place where there are no limits, where you can be anything, do anything, harm anything, gain anything or conquer anything. You must understand this. Even in a sparsely inhabited region of space, you would not be able to do this for long.

Your neighbors are powerful. You do not want to fight against them. It will be enough to protect the borders of your world from inappropriate intrusion—from resource explorers and economic collectives, from opportunistic groups who want to establish influence here and who want to establish trade here. To protect your borders will be difficult enough. No one is going to come and take the world by force unless you prove to be so aggressive, so destructive and so destabilizing to the local region of space that you prove to be too great a security problem for other worlds.

However, the possibility for this is long into the future and humanity has great problems here to face at home—problems which will determine its ability to unite and to remain viable stewards of this world. You have the Great Waves of change to face and all that this will require of your nations and your peoples. You have to offset the Intervention that has been occurring in the world today, that is being carried out by resource explorers—by groups that are not well respected in this region of space, by those who travel around trying to take advantage of whomever and whatever they can. These groups do not represent established worlds or established powers, for these powers do not conduct their affairs in this manner.

You can see here how humanity has not established its collective maturity and has not reined in its destructive tendencies and its ambitious behaviors, its greed and so forth. You can see here that you are like a wild teenager, but you live in a neighborhood of established adults. Yet there are predatory forces in your local universe who will try to take advantage of your recklessness and your naiveté to their own advantage. We shall speak of this in later teachings.

For now, you must come to accept that there are severe limitations to what humanity can do in your local universe. As you grow and expand, if you are able to withstand the Great Waves of change that are occurring in your world, you will have to contend with other races who have been established for a very long period of time and who have established rules of engagement, rules of travel and rules of trade that have taken a very long time to establish and that are well fixed.

You have to grow up and become responsible and, if you can, maintain as much self-sufficiency in this world as possible. For if you lose your self-sufficiency, you lose the great possibility and the promise of being a free and independent race. To become part of larger networks of trade, you will be influenced by these networks, and they will determine to a great degree what you can do and what you cannot do. You will have to pay a very high price for engaging in trade with them, for they will influence you and, in certain cases, try to take advantage of you. You represent a newcomer—a weak and unstable race, a race that is easily persuaded and manipulated. For even very stable nations will seek to take advantage of weak emerging races such as your own for their own benefit and stability.

Such is the nature of life in the physical reality. Whether you are a native living in the world or whether you are living in an advanced nation in the universe, it is the same reality. It is the same problem of competition and survival, of persuasion and influence. The rigors and the difficulties of manifest life do not end by cultivating advanced technology. In fact, cultivating advanced technology complicates your life, makes it in many ways more difficult and more challenging and invites the interest and influence of others who you must then contend with. This is a fact of life throughout the universe. It is a reality of physical life itself.

The wealth that you have you do not want to flaunt. You do not want to show off in the Greater Community. You will have to learn to be extremely discreet about whom you engage with, the nature of that engagement, who can visit your world, what they can do here and so forth—all things you are not doing now that you will have to do in the future.

It is like growing up as an individual person—young and reckless, thinking you can be anything, have anything and do anything, being out of control and unaccountable. You then enter a mature environment, and you begin to face the kinds of restraints that exist there. You find out that not everyone is interested in you, not everyone likes you and you cannot do whatever you want. To accomplish anything takes a great deal of planning and work. If you want to be an honest and happy individual, you must restrain certain kinds of passions and tendencies within yourself. You must become aware of certain seductions in your environment that can undermine your integrity and weaken you as a person.

The analogy here is very strong. It is very appropriate. The hazards that a young adult person faces in the world—the seductions they will face, the influences they will face, the restraints they will have to deal with—are very much like humanity's position in space now. You will have to face restraints, influences, seductions and difficulties. And yet, as growing up for a young person is necessary, important and holds some great advantages and opportunities, so growing up as a human family holds great advantages and great opportunities.

To establish unity in this world and to end tribal conflict will be an immense accomplishment for humanity, allowing you to devote your energies to more productive and creative endeavors. Being a part of the Greater Community, you will learn wisdom from the Greater Community, wisdom that will enhance you both on an individual level and on a collective level. For while there is much competition and freedom is rare in the universe, there is much wisdom there. The New Message for humanity is bringing wisdom from the Greater Community to you in such a way that you can understand it and apply it within your own individual life.

In understanding that all races must seek resources, you can begin to understand why anyone would come to your world. They either want to enhance their security, or they are here to gain and acquire resources. There is no other reason to go visit worlds. Do not think there is a lot of tourism in the Greater Community. Perhaps between trading nations there may be. It is the problem of biological hazards. You may travel to another world to visit a museum or some kind of natural feature of that world, but because you are not of that world, there is always the problem of biological contamination. And this problem has not been universally solved. Between nations or worlds that have regular contact, it can be controlled. But for an outsider to enter these realms would either be prohibited or extremely restricted. So you must understand that anyone who comes to your world comes either to enhance their security or to gain present or future access to resources.

In the Greater Community, you are dealing with an adult environment where the need for resources is ever present and is extremely pressing. No one is going to come to this world on holiday. No one is going to come to this world for a science project, leading a classroom of students as on a field trip. This world is far too hazardous for any foreign nation to do that.

You can see here already the degree to which your assumptions and expectations are out of keeping with the reality of the situation, how your expectations regarding contact with extraterrestrial life in the universe must really be based upon a different foundation and that there are criteria that must be created in order for this contact to be really beneficial to everyone involved.

You must then look at the stars not as a place of escape and not as a wondrous environment where you will be free of the difficulties and restraints of life here on Earth. You must look at the stars and space not as a refuge, but as a representation of life on a larger scale.

Much of what you face here on Earth you will face in space, but on a much larger scale and in more extreme ways.

For example, the lack of freedom that exists in many parts of the world will be even more extreme in the Greater Community. The strenuous struggle for resources that you experience in the world, particularly amongst poorer people, will be more extreme in the Greater Community. The need to control societies and individual behavior is more extreme in the Greater Community. The problems in trade, travel and negotiations are more extreme and difficult in the Greater Community. The necessity to restrain war and conflict is more extreme in the Greater Community. The Greater Community is a bigger version of your life here, bigger in all ways.

The problem of maintaining physical health is more extreme in the Greater Community, where you are confronting beings who have evolved in completely different biological environments or who have evolved in sterile environments. The problem of contamination is extreme. Someone could come to this world, contract a virus or a set of viruses, take them home and infect their entire planet, wiping out most if not all of the inhabitants of that world. That is how potent the problem of contamination really is. Advanced medical technology has not erased this risk.

The care that one must bring to one's life and affairs is more extreme in the Greater Community. The problem of getting around is more difficult in the Greater Community. The problem of getting along with others who are different from you is more extreme in the Greater Community because the differences between you and them are so much greater. The problem of communicating to others is more extreme in the Greater Community, where differences in nature, temperament and appearance are so great.

Therefore, do not look to the stars as an escape from life. Do not look at the prospect of space travel as a wonderful adventure into uncharted and uninhabited regions. Do not think that you can go anywhere you want and that you will develop over time a technology for traveling quickly through all the dimensions of life, zipping around the universe as if it were your own neighborhood. Do not think your conveyances will take you anywhere you want in the blink of an eye. Such notions are to be expected from young races full of imagination and hopeful expectation. But they do not fit into the reality of life as it has evolved over a much longer period of time.

There are nations in your local universe that have been in existence for twenty or thirty thousand years. They have established a level of conformity that is very strict. They are not engaged in social innovation, and most of them have no notion of individual freedom whatsoever. You would find living within them to be extremely difficult, if not intolerable.

Free nations are rare and very distinct. They do not travel around trying to plant their flags in other worlds. If they engage in planetary commerce, it is to a very limited degree. And they do not welcome visitors to their worlds. They remain isolated and discreet and guard their boundaries very carefully.

As We said at the outset, much of this revelation will be different from your ideas and perhaps very disappointing to your expectations, even to the point where you will not want to accept what is being presented here. You may still want to have your hopes and your dreams and your fantasies. But, alas, the universe represents all the problems and the difficulties of physical life, the problems and difficulties of living in separation—in a separated state, separate from God. These problems exist here and everywhere in the physical reality. No one has been able to escape them entirely. Even races that have established a very stable self-sufficiency, even they have real problems in governance and in dealing with outside forces. And they must be very vigilant regarding their encounters and the influences they might receive from the Greater Community itself.

So emerging into a Greater Community of intelligent life, you are emerging into a vast and complex demonstration of life that is in many ways similar to what is required of you here on Earth. What is different is that there are many different participants who are very different from one another and very different from you. And the complexity of engagement, trade and commerce is great and requires tremendous sophistication, great care and great vigilance. Yet the possibilities for one's own world can be significant, if one proceeds without aggression and ambition and if one recognizes that freedom and stability are of far greater value in the universe than conquest or expansion.

Civilizations and nations that have been able to survive over time have come to these conclusions. Those that seek to expand and conquer have found themselves in the end destroyed or overtaken by others, for their behavior could not be sustained in a Greater Community environment, where there are many races seeking to maintain stability and security.

The greatest threats to even the most established nations in the universe are: the loss of resources, biological contamination and environmental collapse. These three things affect the behavior of stable and evolved nations more than anything else. While war and conflict are rare in many regions of space, competition and influence are abundant. And this threatens the sovereignty of nations and their access to necessary resources.

Therefore, what concerns the established nations in your neighborhood in space are different from what preoccupies the nations of Earth who are focused on growth and expansion in a world of declining resources. Your circumstances will require you to mature or to fail. Stability and security will become ever more the emphasis of your peoples, your nations and your leaders. The idea of conquest will become ever more dangerous, destructive and counterproductive in your affairs with one another. And the need to secure your resources will become the predominating focus of this century and of the times to come. It is because you are emerging into a Greater Community and because you have grown in the world to such a degree that this emphasis now becomes your focal point.

Unbeknownst to you, you are preparing for the Greater Community. Facing a world of declining resources, you are actually preparing for the Greater Community, where the problem of securing resources will be continuous and ongoing and is being universally experienced. Real freedom, if it can be established and maintained, will require unity, self-sufficiency and great discretion—here and everywhere.

Though in the history of your local universe
there have been wars and great conflicts,
in the last ten to twelve thousand years
there has been a period of great stability.

CHAPTER 4

Trade and Commerce in This Region of Space

In areas of space where there is a large concentration of advanced nations, one will usually find a great deal of trade going on. This trade would be carried on through established trade routes that are marked in space and are delineated by coordinates with other known planets and systems.

In the region of space where Earth exists, where there is a large concentration of trading nations, there are several major routes and many minor routes that are used for trade. Some are open for all travel, and some are private and can only be used by certain groups who maintain these routes for themselves. Large routes are usually protected by a security force that is generated by the overseeing body that manages that particular route. Sometimes routes are maintained by one nation and sometimes by a whole assembly of nations. Very large trade routes in the Greater Community are managed by large overseeing organizations who provide travel protection for all of their member or client states and nations.

Trade is essential for advanced societies. As races gain greater technological advantages and as their populations increase, they usually outstrip their world's natural resources and must trade and explore for the fundamental things that they need. The greater the technological development, the greater the need for resources. Many resources used in advanced forms of technology are quite rare and must be gained from afar through a complex system of trade and commerce. Many materials are brought from outside a region where a nation usually functions. Negotiations and contracts can be complicated, as there are many competitors for these kinds of resources.

The extent to which a nation must trade will compromise its self-determination and in some cases even undermine its sovereignty entirely. It is common for young, emerging races such as your own to fall under the domination of foreign powers as a result of becoming dependent upon certain forms of foreign technology. In some cases, this domination is outright and complete. Yet often this domination is exercised in such a way that the native peoples of a world will continue to think they are self-determined and sovereign when, in fact, their race has become entirely dependent on foreign powers for the essential things that they have come to rely upon.

Within free nations in the universe, and particularly in large inhabited areas such as the region in which your world exists, great effort is made to sustain self-sufficiency or to participate with one or two other free nations to create a network of support. Here trade is carried on within this network almost exclusively. This is done to preserve a free nation's self-determination and sovereignty and to limit access to other powers who would seek to gain trade and influence within that nation.

Privacy is very important in the Greater Community, and the degree to which one must trade will determine how much privacy one can enjoy and expect. If you have great wealth, either by living in a world with great biological diversity or by possessing resources that are desired or needed by others, then it becomes extremely difficult to maintain privacy and to limit the influences which will be exerted upon you continuously by other nations.

In the region of space where your world exists, conquest is not allowed. It is strictly forbidden, and this rule is maintained in order to assure order and stability within this neighborhood of life. If one race seeks to gain advantage in another inhabited world, it must do so in such a way that it appears that its presence is welcome in that world and that a mutual agreement has been established.

This is very important to understand in your world, where intervening forces are already here attempting to establish influence. Should it appear to outside observers that their presence is at least tolerated, if not welcomed, then there will be no effort on the part of other nations to restrain their presence here. But should humanity exercise its authority and proclaim that it does not seek intervention and does not welcome those races who are present in the world today, then those forces must withdraw or face considerable difficulties from their competitors and from other nations in the region. Conquest leads to instability and insecurity. It demonstrates aggression, and aggression in an established region such as this is not tolerated at all.

Within the realm of trade and commerce, there is the established legal trade dealing with resources, technology or information that participating nations are known to have. If this trade is carried on in areas where there is a small network of participating nations, then what they establish to be legal or beneficial is up to them. Yet in large trade routes where many nations—even hundreds of nations—trade, the rules are established as to what is appropriate and what is not.

However, in all cases, illegal trade is also carried on, and great effort is made to conceal this trade. Sometimes it is smuggled within established trade routes, and, as is often the case, secondary routes—where no official protection or security are provided—are used to transport goods that are technically illegal within a certain region or district.

While stability and security are maintained as the basic focal point in this region of space, nations still seek advantage over one another. The desire for wealth is universal. This, of course, generates illegal trade regarding things that are deemed socially inappropriate or that are considered hazardous to the security and stability of a region. Dangerous drugs, certain biological elements, slavery, slavery trade and things like this are generally included in the list of items that are considered illegal and inappropriate.

Free nations that are self-sufficient or that maintain their own trade networks can remain relatively anonymous as long as they do not demonstrate that they possess any unique wealth or advantage. They are thus free from many of the difficulties involved in illegal trade and in piracy, which occur even in established regions and which are certainly a problem in outlying districts and in territories that remain uncharted.

Trade and commerce are both a benefit and a hazard. They provide a nation with access to resources that its world or collection of worlds does not possess. Yet they carry great hazards because wherever there is commerce, there is influence, and wherever there is influence, there is competition. Here one must gain great sophistication and establish very clear and sustained rules of engagement regarding commerce with other nations. And there must always be great care that the risk of contamination is kept to a minimum or is avoided altogether.

In trade and commerce, there are always attempts to discover the inner workings of nations through the avenues of trade. Espionage is not uncommon and has achieved very sophisticated expressions. Persuading individuals in other nations to divulge information or to provide access to opportunities or to other networks is continuous between trading nations.

So while nations may not be at war with one another, they are certainly competing, and if one nation has an advantage, all the others want to know what it is and how it can be counteracted. Tremendous resources are brought to bear in this matter. It is so much like your world, except that it is happening on such a greater scale, with many more participants, and has taken on far more advanced and sophisticated demonstrations.

If a nation loses its self-sufficiency and must seek abroad for its basic requirements, such as biological resources, basic technology, food and essential medicines, it will be very difficult for that nation to remain independent and self-determined. It will have to yield to the terms that are provided, and this can be a very difficult situation. As the need for resources is everywhere—the need for food supplies, for biological materials, for materials for the development of medicines, for metals, for information and for many other things—it can be very difficult for a small nation to acquire what it needs without making serious concessions.

All nations that are involved in contact in the Greater Community have their secrets, of course, and some protect their proprietary information with great effort and determination. Yet this is difficult to do in the Greater Community for reasons which will be discussed through the series of teachings provided here. Everyone has their secrets and everyone is trying to know each other's secrets. So while there can be cooperation between nations and sometimes even a close interdependent bond between nations or amongst a group of nations, the problems of secrecy and the problems of unequal wealth continue.

Many nations in the Greater Community have united with others to create networks of trade. But even within these networks there are problems of trust, problems regarding espionage and certainly problems in dealing with the influence of outside forces. It is most often the case that the larger the network, the more difficult it will be for individual participating nations to keep their secrets and proprietary information. Small nations that have minimal contact with the Greater Community fare much better.

Stability and security between nations require maintaining a continuous access to resources and a careful guarding of any information, technology or unique resources that one world or a group of worlds may possess. Those who have wealth here have a great disadvantage. If they possess resources that are greatly needed by other nations, it will be very difficult for them to protect these resources and to maintain their privacy. Their wealth will be noted, information will be shared and many will be seeking to gain advantage or influence there.

In all, it is a very complex situation. Strict rules of conduct have been established in order to maintain stability and to avoid any breakdown in the functioning of trade and commerce. These rules are often maintained by courts and are supported by security forces with an emphasis on honesty. Here corruption is certainly a problem. Involving trade between many participating nations, corruption becomes counterproductive if one is attempting to establish security and stability in a large region.

If your behavior becomes unethical or untenable, others will not engage with you in commerce. You then become isolated. The risk of isolation for a trading nation is an immense threat, a threat that is usually sufficient to convince that nation to behave in alliance and accordance with the wishes of others. That is why losing your self-sufficiency is such a great disadvantage. You will have to conform to the expectations of others and to the rules of commerce if you are in such a position.

As no one is perfect in the physical universe, there of course is crime and deception. Some of these problems are local while others are manifest everywhere to some degree. Competition can breed innovation, but it also breeds deception. And deception is carried out to a great degree within the boundaries and the rules that are established between trading nations. This has led to some amazing creations and many methods in determining the honesty and sincerity of others. Even technology has been established to determine the truthfulness of one's statements, one's contracts and one's declarations. As lying and deception are dangerous, they have been counteracted in many, many ways.

This has led to great stability, as it becomes counterproductive for any trading nation to try to deceive its partners because deception has become easier to detect and the threat of isolation is always there. A highly advanced technological society would fail quickly without constant inputs of technology and resources. In your district of space, this has created, through persuasion and the threat of exclusion, a relatively stable system—a system that is generally followed by its participants. Over time, this has been established. Though in the history of your local universe there have been wars and great conflicts, in the last ten to twelve thousand years there has been a period of great stability, and this has been maintained with tremendous determination.

It is in humanity's interest, therefore, to maintain its self-sufficiency, to restrain its ambitions and to be grateful for what it possesses, without wanting ever-greater sources of power and influence. This will provide humanity greater anonymity, security and freedom from intervention and from forms of persuasion that you have not yet learned how to counteract. Because you live in a beautiful world that is desired by others, this will be required of you if you seek to remain a free and sovereign people in the universe.

Do not, then, be desirous of things you do not possess. Do not seek to have all the advantages that others seem to have, for these come at a tremendous price. Do not seek wealth beyond what your world and your solar system can provide, or you will fall prey to the persuasions that exist in the Greater Community, and you will be facing competitors that are long established and who have far greater powers of persuasion, skill and technology than you do.

The Greater Community is an environment that requires tremendous discernment and discretion. If you are to participate here in such a way that your own privacy and freedom may be sustained, then you must develop this discernment and this discretion. You have native skills that God has given you to do this, but they must now be brought to bear. Humanity will have to unite to accomplish these things and to end its ceaseless conflicts amongst its tribes and nations in order to prepare itself to function in the Greater Community. This will be born of necessity. It will be required.

Therefore, do not seek for foreign technology, do not invite it and do not accept it if it is offered to you. For if it is offered freely, it will not be offered by a friend but instead by a competitor in order to trap you, to make you dependent, to make you want more, to make you desirous of things that only others can provide. No true ally of humanity would offer humanity technology, and that is why whoever does, does this for other purposes. Do not be seduced, and do not be induced to engage in trade of this nature, or you will unknowingly and unwittingly begin an engagement that will be very difficult to disengage from later on. That is why restraint is so important—wisdom and restraint.

For a world such as your own, extensive trade and commerce would destroy human culture. It would undermine everything that you have that is noble and great, and you would become involved in a complex situation that would require a great deal of your focus and energy to sustain. That is why there is a great truth in the Greater Community that the wise remain hidden. They remain hidden to remain wise, and they remain hidden to remain free. This is an important lesson and an important truth that has been established throughout the universe, a truth that humanity has not yet learned to recognize or to value.

Distant trade is often not conducted because of the dangers and hazards involved. To move beyond the boundaries of a well-inhabited district means that one must travel through unknown and often hostile territories. The risk of becoming lost, the risk of piracy and the risk of inviting unwanted scrutiny are significant. Even in regions where there is very little intelligent life, in regions where intelligent life has not been cultivated, the physiological hazards are significant.

This is why uncharted territories often remain uncharted. They are often beyond the reach of traveling nations. Explorers do venture into them but often do not return. There are nations within uncharted territories that seek to remain entirely hidden from the outside and have established very complex and successful means of discouraging exploration and intervention.

Those few free races in your local universe who are aware of you, who value you and who wish to see you emerge as a free and sovereign people will not encourage trade or commerce for you. And they will generally not seek to establish trade and commerce with you unless you deplete your world to such a degree that you simply cannot sustain yourself here. It is better that you remain technologically limited, but free and self-determined, than it would be for you to become more technologically advanced and to fall under the domination and control of foreign powers in the Greater Community.

The assumptions that prevail in the human community at this time are extremely ignorant of these things. The desire for technology, the desire for resources, the desire for contact with the Greater Community—it is natural that you have these expectations, but they do not represent real wisdom on your part.

Your true allies in the universe will not seduce you with technology, nor will they induce you to become involved in trade and commerce, for they know the great risks and the price that this requires. They will encourage you to remain self- sufficient and united and to develop a very clear and sustained boundary to space.

Because you live in a relatively stable region of space, you do not need a great military power, but you will need to maintain sufficient balance and order within your own world, and you will need to maintain a very clear and determined control of your solar system. Others races accept that your solar system represents your sphere of influence. This is common and is expected. It is very rare that two powers can exist within one solar system for very long. Either one will destroy the other, or they will both destroy one another in competition for dominance there.

Your solar system will provide certain degrees of wealth and advantage for you. But if you destroy the life-sustaining resources on the Earth, if you drive your climate systems into a different kind of functional stability that creates an inhospitable environment for life, you will not be able to escape this within your solar system. You will not be able to find reprieve or refuge.

If you understood your position in the universe and the difficulties and requirements that exist in this part of the Greater Community, you would consider your world very differently, and you would behave very differently. You would not squander your wealth or your resources on conflicts between your groups, nations and tribes. You would have an entirely different set of priorities.

This instruction regarding trade and commerce in the Greater Community is very important for your future. There will be many attempts to induce you to become involved in receiving gifts of technology. There are many commercial forces who would seek to have you as clients and to gain control and access to many of the important resources of your world. The inducements will be great. The promises of wealth, peace and power will be great. And who amongst you—amongst your leaders, amongst the wealthy people of the world—could turn down such offers and resist such inducements?

This is why great wisdom and restraint must be established here. This is why human freedom and sovereignty within this world must always be a great consideration. While others in your neighborhood of space are not free to dominate your world through conquest and through military force, they are entirely free to gain influence here under certain conditions. And they would not hesitate to want a world such as yours as part of their network, as one of their primary clients or as a resource within their sphere of influence.

The wise remain hidden. Your world is known to a few, but not to many. You do not want to broadcast into the universe who you are and what you have. In the future, your radio technology will have to be completely changed. You cannot be broadcasting out into space. That only invites inquiry, interest and suspicion. Right now you think that there is no one else out there, that the universe is vast and empty and that there might be some distant planet somewhere that could receive your transmissions. If you understood the situation, you would see how dangerous and unwarranted this is.

Remember the three requirements for freedom in the universe: unity, self-sufficiency and great discretion. Practice this and you will be able to enjoy the benefits of emerging into the Greater Community as a free and sovereign race and to gain a unity and stability the world has never had. Humanity has been growing and expanding in its evolution. But it must reach a sustainable state—a state that functions at the level of stability and security, that is not oriented towards growth and expansion. If you seek to grow and to expand out into the Greater Community beyond your sphere of influence within this solar system, you will run into tremendous difficulties. You do not want to be seen as an aggressive race by your neighbors in this part of the Greater Community. They will not tolerate such behavior, and they have the power to suppress any attempt on your part to gain territory beyond this solar system. Nor do you want to enter into competition with them in a hostile or contrary manner, for this would not serve your future in any way whatsoever.

It is not yet certain whether humanity can sustain this kind of wisdom and restraint. Even now you have very little else to conquer in your own world. You have your solar system to explore, but you will find very few places where you can make any establishments at all, and even these will be difficult to maintain. You will have to face and are facing even now the pressure of influence and persuasion created by opportunistic groups in your region of space who are here to do business, who want to extend their power and influence through means that are within the bounds of what is tolerated.

Dealing with this successfully will elevate your race and give you a greater maturity. It will certainly bring an end to human conflict as you have known it in the past. For you can no longer afford to waste your resources through war and conquest. You must now prepare to develop a steady and sustainable state that you can maintain within this world without dependence on foreign powers. Neglect this and you will face the same fate as so many nations who have aggressively expanded only to find themselves completely at the mercy of others. These others will not share your values, for they will not be human. They will not value your unique gifts and capabilities. They will only see you as a security problem.

This is a harsh reality, but one that gives you scope and wisdom. There are certain requirements you must achieve if you are to remain free and self-determined. You cannot change these rules. They are rules of life. They are the requirements of living in the Greater Community into which you are now emerging.

The Greater Community is your future.
It is your destiny.
But you must be prepared.

CHAPTER 5

TECHNOLOGY AND THE DIFFERENT PATHS TO STABILITY

The Greater Community that you will experience has many restraints. It also has many opportunities. It requires that nations become united within themselves if they are to engage with other worlds and with the difficulties and complexities of Contact. This is a great advantage as long as it is carried out in a way that is truly beneficial for the peoples of that particular world. This requires unity, a unity that is not born of one philosophy or religion but of shared necessity.

Nations within your world now must work together for the security of the world. Here again, stability and security become the overriding emphasis. Ancient animosities and hatred between nations and tribes must all be moderated so as not to break out into war or destroy the world's resources.

Here you become like the Greater Community you will face. You will have to suppress certain tendencies within your nature and within your world in order to achieve stability and security. And you will have to develop a boundary between your world and the Greater Community in order to exercise your own rules of engagement and to determine what ethics you will follow regarding who you will be in contact with and how you will engage with them.

In the Greater Community, in the region in which your world

exists, nations usually engage with each other in space, in counsels, and rarely on their home planets. Because secrecy and discretion are important, actual visitation to other worlds is rare unless worlds are in a network together and have developed over time a great trust with one another. In this case, many different races, if they can tolerate a similar environment and environmental requirements, can live on several different planets. But actually multiracial environments are fairly rare because of the biological hazards involved. Unless races have evolved to function together and have developed the necessary technology and medical boundaries to prevent contamination, then you rarely find many different races living in one world. Yet if races live constantly in sterile environments, cohabitation is possible and is practiced.

You can begin to see the limitations here. It is rare that you will find areas on the terrestrial surface of planets where many different racial groups function and travel, come and go and so forth. Trade is conducted generally off planet through large trade stations and networks in space. Many nations will have trading stations near their planet, or planets, in order to control exposure and to ward off unwanted inquiry and intrusion.

Here you can begin to see the great problem of keeping the Greater Community out of your sphere of influence. If you allow it into your sphere of influence, then you will face tremendous difficulties, and stability and security will be more difficult to sustain and, in some cases, impossible to maintain.

Societies evolve and build within these restraints if they are to survive and pass through the many thresholds of development, both within their worlds and in contact with other worlds. Eventually, they will generally establish these kinds of parameters. If their worlds have reached a peak in what they can sustain for their own populations, then stability and security become the emphasis. Outside interference is recognized as hazardous. Trade, if it is engaged in, is dealt with in a very specific manner. Societies evolve around these restraints.

While it is necessary for a world to become united within itself to function successfully in the Greater Community, the nature of that unity can take a broad range of expressions. To be a free race, to be a race that has individual rights and freedoms within its own world and sphere of influence, requires a great deal of development and restraint. As worlds become overpopulated, as resources become depleted, the restraint on personal freedom increases.

This is what you will be facing now in the decades to come—the loss of wealth, the loss of mobility, the loss of physical space, the loss of opportunity and ever-greater restraints imposed upon you by your own governments. This is what happens when you reach a point where you are overwhelming your world's resources, where you have run out of room and where you are at risk of undermining your world's ability to sustain you in the future.

That is why humanity at this moment is creating the very conditions that will limit or even destroy individual freedom here. This is a very great problem. It is a great problem that will be determined by how you—as part of the human family, as part of a nation and even on the individual level—will deal with the Great Waves of change that are coming to your world. The Great Waves of change are largely the product of the overuse and misuse of your world's resources, the pollution of your world and the great instability that is being created in the world's biological, ecological and atmospheric systems.

There are worlds that emerge into the Greater Community

where personal freedom has never been known or practiced. But, generally, worlds such as your own that are biologically rich and where the native peoples evolved in isolation from one another go through a very prolonged and difficult process of establishing contact and unity. This is rarely done in a harmonious manner, as competition and conflict overwhelm people within a world where certain groups become dominant and where wealth is not shared or distributed equally.

The pathway, therefore, that humanity has followed in its prolonged and very difficult and unhappy history is actually a pathway that is followed by many other emerging worlds where intelligent life has been able to evolve and to sustain itself. Do not think, then, that humanity is a bad race and that humanity is more evil or sinful than other races. Actually you have followed a very normal path of development in the Greater Community thus far. But the limits of time and the requirements of your environmental situation in your own world now require a great change—a great change in emphasis, a great change in understanding, a movement away from growth and expansion to security and stability, to sustainability, to more of a steady state, to a greater equality and equalization between nations and peoples.

Here you will have to control your population size. You will have to control the distribution of wealth. You will have to provide for the poor. And the wealthy will have to moderate their behavior, their appetites, their greed and their consumption to achieve stability and security.

Technology will have to advance, but even more than this, you will have to change the way that you live. Those who now consume too much will have to consume less. And those who have too little will have to have a sufficient livelihood. Over time, the human population will have to become smaller, hopefully through very humane means and through restraint, for there will no longer be any room for an ever-growing human population. You will have to live within greater restraints.

Other worlds have had to follow similar paths. However, in many cases, worlds were overtaken before they ever became united. Foreign influence became established, inroads were made into the leadership of emerging worlds, their conflicts were exacerbated until one chosen nation or group was supported to overcome all the rest. This has not yet happened in your world, and it is to your benefit that this is so. Some worlds are so valuable to others that they were overtaken before the native peoples ever gained any technology. This has happened countless times in the Greater Community.

In the stable environment that exists in your region of space, your world is amongst a very few that have not been overtaken or consumed by other nations. The biological diversity of your world has played a part in this. The value of your world has restrained other nations from entering into direct competition with one another. But now, amongst those races who seek to gain dominance here, there is a competition, a competition for influence—a kind of subtle attempt at conquest to make humanity weak and dependent, to encourage human conflict secretly so that humanity never gains security or stability, thus requiring it to reach out to foreign powers and to accept whatever inducements these foreign powers might offer you.

Your few allies in this region of space speak out against this, as they have done in the Briefings the allies of humanity have sent into the world. But intervention and influence are part of the problem and the challenge of emerging into a Greater Community of intelligent life. Yet this is far different from most people's expectations and assumptions.

Generally speaking, if nations become technologically advanced, they become entirely secular in nature and will rarely have active, vital religious traditions. This is because the emphasis on technology and resources becomes so predominating and such a source of wealth and control that it overwhelms whatever other notions of power and authority that the native peoples might have had. As a result, over time, the reality and the existence of religion as you know it becomes very rare and is only practiced by small groups, often under clandestine conditions. Therefore, you should never assume that technological advancement promotes ethical or spiritual advancement, for in most cases it is the opposite that actually transpires.

Great technological powers in your region rarely allow any individual freedom amongst their peoples. They function with great uniformity. Their citizens are expected to conform to very strict patterns of behavior to maintain this emphasis on stability and security. This emphasis has very unfortunate outcomes here for races that have lost their spiritual foundation and that have lost the emphasis on individual freedom and creativity.

There are some advanced races that have been able to maintain this freedom and creativity to their advantage and to the advantage of their peoples. But, generally speaking, technological societies, particularly if they have extended their sphere of influence beyond their solar systems, tend to have very rigid social structures and, as a result, have created environments that you would not find conducive to life—environments that you would have great difficulty living in.

The emphasis for your world, then, is to achieve stability and security in such a way that individual freedom is preserved and honored and that individual creativity is valued and is put to good service in the world. This requires a very unique kind of development.

Therefore, do not be in love with technology, for if you are, you will fall prey to those who are more technologically advanced than you are. Do not think technology will assure human freedom and well-being, for technology is a power that can be used for good or ill. This power must be restrained and coordinated with other priorities—keeping Knowledge and the spirit of humanity alive through the difficult times ahead, through the Great Waves of change. Only this will preserve human freedom in a world that will become much more ordered and restrained.

Do not be impressed by races that have advanced technology, for most of them have had to sacrifice things of great value to achieve this technology. Their quest for power has cost them freedom and, in some cases, their race's self-determination. You will find that in most advanced technological societies, there is very little freedom, very little emphasis on individuality and the importance of individual creativity and so forth. Here people are considered in terms of groups, not as individuals. They are rated according to their groups' abilities and not according to their individual talents. They are seen as part of a functioning society—a society that requires a rigid order to be able to function, a society that has overextended itself in the Greater Community and that has destroyed and overwhelmed its natural resources. These societies become uniform and oppressive. Even if they conduct peaceful relations with other nations, their own citizens are forced to live under tremendous restraint and expectations that are counterproductive to the individual's freedom and well-being.

Unfortunately, the ascension of power in the universe generally leads in this direction. Those races who remain free in their evolutionary progress must therefore accept limits to their technology and their desire for technology, realizing that technology is but one part of what makes life possible and sustainable.

Even today in the world you can see the emphasis on technology, the belief in technology, the belief that technology will solve all problems and the reality that technology is becoming ever more the focus of people's emphasis, as if it were a religion of its own.

Even amongst your nations, the wealthy nations in particular, you can see the emphasis on the worship of technology, the belief in technology, the belief that technology will save you and that technology is the real emphasis in life. Many young people in your world today have a stronger relationship with their machines than they do with any other person. This is moving in the same direction that so many races in the Greater Community have followed, and the outcome is unfortunate.

Whatever you are in relationship with influences you, and you become more like it. If your primary relationship is with technology, you become more machine-like yourself—thinking only along certain corridors of understanding, basing your information only on certain precepts, controlling your mind, limiting your mind, ignoring your greater powers and your greater insights. Rational self-control and human logic become very machine-like in this regard. This conformity and rigidity is highly promoted amongst most advanced technological races in the universe.

Already humanity is falling prey to this seduction, and a seduction it is indeed. While you will need to develop your technology to establish stability in your own world and to provide food and resources for your growing population, you must always remember that it is the greater power of Knowledge within the individual and within people together that represents your core strength, not your technological achievements. Never lose sight of this, for if you do, you will lose that which is most valuable to your life, to your future and to your self-realization.

It is natural that races who are intelligent will evolve more complex social systems and eventually will assume certain forms of technology to give them advantages within their own native environments. But the quest for power is seductive, and beyond meeting practical necessities, it becomes an emphasis in and of itself. Here restraint is important. In the Greater Community, should you establish a technology that is unique or that others do not have, you will become the focus of inquiry, influence and even intervention.

Social structures either move towards a healthy equilibrium or a restrained equilibrium in order to achieve stability and security. Here there is either an emphasis on individual freedom and the value of the individual, or there is not. The process of building stability and security requires that people find ways to work together and to cooperate together. This is either forced upon them or it is something that they create for their own benefit. Choosing which path to follow represents one of the great thresholds in the evolution of a race.

Humanity will need greater stability and security now. This need will grow immensely. Whether humanity chooses to impose order or to cooperate to create order remains to be seen. Whether humanity will achieve stability through a greater consensus or whether stability will be forced upon people from a hierarchy of power remains to be seen. Whether people can live with greater restraint willingly and respectfully remains to be seen within your own world.

Too often in the Greater Community, stability and security are established by the dominance of a governing power, and this is always the case when nations are overtaken by other nations or become economically dominated by other nations. The suppression of the individual is always the result.

Therefore, freedom is rare in the universe, and freedom is the most important thing. The freedom to be a self-sufficient, self-determined race of people and the freedom of individuals to contribute their gifts to their society and to their world are the greatest emphasis, not the assumption of technological power.

Humanity, in its long and difficult history, has with few exceptions always chosen the path of dominance and power. If you follow this course, it will lead to an inevitable end, an end that has been demonstrated repeatedly and consistently throughout the Greater Community, where freedom is rare. You can learn from other worlds, and you can learn from the demonstration of this within your own world and history.

In highly structured societies, conformity was required and has persisted. They achieved stability and security to a greater degree than other societies, but at what cost? These societies did not advance in any way other than in their technological development. As a result, these societies are miserable environments to live in unless you are at the top, unless you control the power. They are extremely oppressive and destructive to their populations.

You will have to find a path to stability that is not oppressive and destructive. This will be perhaps the greatest challenge, along with preparing for the Greater Community, that humanity now faces or has ever faced before. Yet it is this difficult set of requirements and circumstances that will give humanity its greatest opportunity to advance spiritually and ethically and to build a unity and cooperation that have never been achieved in the world before. It is the pathway you choose that will make all of the difference.

Therefore, do not be romantic regarding the Greater Community. Do not imagine that it is full of angelic, lovely beings who are just waiting to help a young and struggling humanity. Do not think that it is filled with marvels and wonders of technology that you will be able to fully enjoy and to obtain for yourself. Do not think you can travel anywhere at will, visit any world you like, go all over the universe without hindrance and travel without problems or difficulty. Here you must face a reality that is real and genuine and that reflects many of the same tendencies and difficulties that you experience here on Earth. Yet the difficulties are greater, the challenges are greater and the opportunities are greater within this immense panorama of life.

The Greater Community is your future. It is your destiny. But you must be prepared for it. You must look at it with great sobriety. And given what is being revealed here, see if you still have a desire for it and inclination for it, born now not of fantasy and hopeful expectation, but of a deeper connection—a deeper connection with life within the world and beyond the world and a deeper sense of humanity's destiny and its possibility of choosing a path of freedom, choosing this greater opportunity.

You cannot completely manipulate
what God has created.

CHAPTER 6

FAMILIES AND GENETIC MANIPULATION

In your region of space amongst advanced nations, the family unit as you know it is still practiced in many places and is always maintained amongst ruling classes. In more primitive worlds, it is certainly nature's pattern. But amongst more advanced nations who rely upon technology and who have a great interest in genetics, the situation begins to change.

Family units and family clans leading to greater tribal units is certainly the evolutionary pattern for intelligent life to evolve. This pattern is followed unless it is interrupted or controlled in some way by external forces.

Over a very long period of time, the emphasis on genetics and genetic control has been extremely widespread with results that have been both beneficial and at times catastrophic. It is certainly catastrophic for those who are bred for certain purposes. This has generated both practical and ethical issues, which vary between organizations and races.

The family unit in more progressive and technological societies has at times been maintained all the way up to the ruling classes. But, in general, to maintain stability and security, unfortunate collective breeding has been generated and cultivated to a very high degree. Worker classes, classes of beings who have specializations, military classes and other groups have been the focus of collective breeding, except in nations where it is considered unethical and impractical.

From the human viewpoint, collective breeding is horrific even though at this moment there is experimentation with genetic manipulation under the guise of improving the health and welfare of people. But this represents a seduction of power and a complete undermining of the reality and the potential of human freedom and creativity.

Races who are involved in exploration and the manipulation of worlds such as yours rely upon genetic breeding. The races that they breed are bred for certain purposes and are often deprived of any individualistic qualities, even to the extent that they have no compassion for one another. They are inbred, and their natural tendencies toward self-preservation are also limited.

The problem of collective breeding coincides with the development of large technological nations that must exert uniformity and control, particularly if they have foreign establishments. To have disorder amongst the population is always a potential and a constant problem, and this problem has in many places been addressed through an emphasis on collective breeding.

But collective breeding has its own hazards, for it is extremely difficult to create the desired results with absolute assurance. There have been cases even in your local districts where classes of bred individuals have turned against their masters, destroying the infrastructure that they were meant to serve. Unable then to fend for themselves because they were not designed for this, they led to great disorder and had to be destroyed.

It is an imperfect science. Even though it has been cultivated to a very, very fine degree of specificity, it is still an imperfect science. For you cannot create what God has created, and you cannot completely manipulate what God has created. A biological entity has other qualities that are beyond the control of genetic manipulation.

So amongst more ethical races, collective breeding has been avoided because it is unethical, because it destroys the capabilities of the individual and because its end results often prove to be unpredictable and even chaotic. Therefore, advanced ethical nations do not engage in this practice.

Yet there are many advanced nations who do not have these ethics and who have made their worker classes function according to very specific genetic designs. They serve those above them who sustain a normal and natural family pattern. Yet this is hazardous, for you cannot control completely the minds and perceptions of those who have come into life, even through the mechanism of collective breeding. The potential for disunity and disloyalty is always there.

Genetic manipulation is a power, but it is a hazardous power. If it is taken too far, if it is employed against the well-being of society itself, then it becomes a great hazard. It is one of the problems of assuming technological power and control and represents the attempt to establish stability and security without an emphasis on individual freedom.

In those races who have engaged in this practice to a high degree, even their ruling classes have become like prisoners, surrounded by a vast army of bred individuals whom they must watch constantly, fearing for their own lives and well-being should any abnormalities or other influences enter into these controlled populations.

That is why the notion of individual freedom and the expressions of individual freedom, such as art, music and other creative expressions, are seen as a threat or a danger to a nation that relies upon collective breeding and upon a vast population of genetically designed individuals. For these notions and expressions might stimulate something in these bred individuals—an interest, a fascination and so forth. Even the simplest expression of individual freedom or pleasure could be seen as a danger to the hierarchy who controls them.

It is an interesting phenomenon that the more one attempts to exert control, the more vulnerable one becomes. There is no state of absolute control and security. Though it is needed to a certain extent to maintain the stability, the survival and the well-being of a nation and a society, if it is taken to extremes, it produces hazards of its own.

God has created the individual to be creative and expressive. If there is an attempt to override this with genetic conditioning, political oppression or social designation, it creates a fundamental conflict within the individual. Here there is always the potential that the individual will turn against its genetic design, will turn against its social designation and will turn against those who have established its functions in society.

Over the course of a very long time, many experiments have been done to attempt to design a hierarchical system of functionality. In very small groups, such as economic Collectives, it has been achieved with a certain degree of success. But in larger established societies, it has become more difficult to achieve. If groups with different genetic designations intermingle with one another, the results can be unpredictable. In larger societies, it is very difficult to separate one population from another entirely. They can stimulate each other in certain ways beyond the predictions and the calculations of their genetic designs.

This is one of the reasons why free nations avoid contact with large nations who rely upon collective breeding and genetic design. It is one of the reasons, likewise, that these large nations avoid contact with free nations, where individual expression and the demonstration of individual creativity would be seen as a danger to the social order by these larger civilizations.

It is a problem certainly in nations that have become entirely secular, where there is no understanding of the spiritual nature of the individual or of the power and presence of Knowledge—the deeper spiritual intelligence within the individual. When nations and civilizations become entirely secular, they lose awareness of the greatest power and potential within the individual. Though they may seek to control the design and functionality of an individual through genetic engineering and social function, they cannot erase completely the deeper spiritual reality within the individual.

This represents one of the great problems facing large, advanced technological societies. It is one of the problems that limits the size of societies and empires. Very large empires are inherently unstable, and the problem of control becomes such that they tend to collapse, either at their perimeters or at their very centers. If you are going to create a society where individuals function like machines, then you have to function like a machine. And machines only function along certain parameters. Outside of those parameters, their nature becomes highly unpredictable. For a biological entity, this will always be the case.

In your local neighborhood, amongst those races whom you may encounter, there is evidence of collective breeding, particularly amongst economic Collectives that carry on local trade and are also involved in illegal trade as well. In larger societies, collective breeding is usually kept to a minimum, for it has proven too difficult to manipulate over time. Yet nations vary from one another in what they create and in their ethical emphasis.

Should you ever have encounters with trading nations in your region, you will see the evidence of collective breeding. You will see it expressed in extreme forms, and you will see the exertion of genetic emphasis in general populations, which will tend to be very uniform in their appearance and behavior.

It is a danger that humanity will have to face because this kind of genetic manipulation is a seduction. It is a seduction of power. Though it may seem to have an innocent purpose at the outset and certain practical applications, the seduction of power is there nonetheless.

Such ideas are now being discovered in the human family. Much of this information has been introduced by those races who are visiting and intervening in the world today. It is in their interest that humanity would develop and use this technology. For economic Collectives have great difficulty dealing with nations of relatively free individuals. It is in their interest that the human populations become more uniform and more like the Collectives themselves.

In family units in the Greater Community, individual freedom is either encouraged or discouraged. In large societies that are technologically based, individual freedom is generally discouraged, and in some places it is altogether unknown. You may serve your family and your family's interest, and there is no deviation from this. You may serve in a certain function or capacity that is designated for you according to your talent, through inheritance or through family status, and there is no deviation from this.

In the Greater Community, the problem with family emphasis

is that it competes with the national emphasis. Families become clans, clans become larger groups and larger groups compete with one another, having no filial connection to one another. That is why the importance of the family in a large technological society is greatly reduced.

Therefore, while you may have offspring in a natural manner, the entire upbringing of your children is completely designated by society at large. Here filial piety and family allegiances are greatly discouraged. In an environment of control, this is certainly the case. In many other nations, there is a relative degree of flexibility regarding family devotion and commitments. It all depends on a nation's structure, emphasis and ethical foundation.

The problem of collective breeding is also demonstrated in the establishment of military forces. Here it must first be emphasized that collective breeding is incredibly expensive, requiring a major infrastructure to educate and to direct an individual's development. If a military body is established through this means, they may be completely loyal to their masters, yet they will be ineffective against an opponent who uses creativity and intuition. They will also be ineffective against an opponent that exerts a power of influence in the mental environment. While a genetically bred individual may be allegiant to its master, it is nearly impossible to genetically breed an individual who is not subject to the powers of influence in the mental environment.

That is why military forces that have been established through collective breeding have proven to be ineffective. They are not creative. They do not adapt well to different environments, and they are subject to persuasion in the mental environment. Great measures have been established, therefore, to counteract military forces that are established and created in this way, to generate confusion and disassociation and to take advantage of the lack of adaptability of such forces.

That is one of the reasons why the development of military power has been limited. It has proven to be very difficult to create and maintain for larger nations. In an environment with a great deal of trade, military aggression is prohibited with great emphasis. Nations have military forces, but they function more as security forces. Only if an entire region was threatened by some external power could these forces be gathered together into a larger and more potent line of defense. Yet in an environment where war does not exist, large military forces would have no purpose. Here the reliance upon weaponry and power in the mental environment becomes ever greater, and the reliance upon defense becomes of the utmost importance.

In the Greater Community, social structures can be very uniform or very complex. In nations where there is genetic breeding, you have very distinct classes of individuals designed for certain kinds of functions. These classes can coexist under the guidance and direction of other social structures that evolved naturally through family units or through collective upbringing.

The family unit experience that you have throughout the world is actually quite rare amongst most large technological societies, who raise children collectively in large groups, largely away from their parents, in order to instill the interests of the nation and to develop individuals according to their talents and the level of their social function and existence.

Only in free societies would you see a social pattern more similar to your own, where an individual is developed according to their specific talents and abilities, rather than simply slotted into a large social pattern for the convenience of the state. In free societies, an individual's talents and abilities are discerned early in their development and are cultivated naturally according to the individual's talents and inclinations. This is done to bring forth the individual's greatest abilities and to cultivate Knowledge within them so they themselves recognize their strengths and their talents and their natural desire to express these in a constructive way. But this form of education is very specific. It is very much focused on the individual. You will rarely ever see this in a large technological society that does not value the capabilities of the individual beyond serving a specific and predesignated social function.

Then, of course, there is training in large technological societies for specific tasks, where an individual will be taken from his or her parents and set into a special kind of education that will completely mold them for a specific social function. You would not see this in a society that is free and that is governed with an emphasis on the individual and their potential and creativity.

This is one of the reasons why free societies are much smaller. They do not have the overbearing social demands of maintaining a large infrastructure. And they do not have foreign establishments that require a very strict conformity where they would expose their members to foreign influence and the powers of persuasion that exist in the Greater Community. Free nations tend to be much smaller and much more inclusive. They must protect their advantages from influences from beyond their world and from the many seductions of power that always exist in engaging with other races that are unlike their own.

Because of the emphasis on individual development, free nations tend to cultivate very high levels of creativity. Advancement in science and technology, in ethics and spirituality, can be remarkably great in these societies. The benefit to society is immense, giving them certain advantages that large technological societies do not have. Yet these advantages invite scrutiny and interest and inevitably the attempts at persuasion. That is why small free nations again must remain extremely discreet. For they will cultivate advantages that larger nations do not have, advantages that will be seen to be of great value and interest to other nations that do not have the same kind of social structure. This again is the problem of having wealth and power in the Greater Community. This again is why the wise remain hidden, and why wise societies remain hidden and extremely discreet.

Here you can see the immense contrast with the human ideal of sharing one's wisdom freely and openly with the universe, of wanting to express oneself everywhere one goes, to share one's gifts wherever one goes and to broadcast one's talents, abilities and accomplishments. It is because humanity has never had to develop a very high degree of discretion and discernment that these motivations and ambitions still exist.

If humanity can evolve into a free nation in the universe, it will have to be extremely discreet, and its greater talents and accomplishments will have to be maintained beyond scrutiny in the Greater Community to as great a degree as possible. Should you develop technological advancements or advancements in your social structure or functioning, even advancements in the arts, much of this will have to remain hidden from the scrutiny of foreign powers. For a free nation is always perceived as a threat to nations that are not free. The expressions of a free nation are always seen as endangering the interests of large nations, where freedom is not practiced and is unknown.

You cannot avoid this reality. And you cannot win over the universe to your point of view or perspective. The attempt to do so will certainly guarantee the loss of your freedom and self-determination over time. That again is why the wise remain hidden to remain wise. The free must function with great care and discernment in the Greater Community to remain free.

Yet this is so different from the attitudes, perspectives and ambitions that are predominant in the human family. The exertion and expression of freedom in your world within the human family are perfectly natural and beneficial. But even here you can see how freedom is viewed as a challenge and a danger amongst nations in your world that are not free or that have greatly limited freedom for their citizens. They are afraid of the expressions of the individual, whether these expressions are beneficial or not. It is a problem that you can see even within your own world. It is a problem of how much freedom you give the individual. In the Greater Community, individual freedom is rare. In many places, it is unknown.

Within your world, as you face the Great Waves of change, individual freedom will be curtailed out of necessity, out of circumstantial necessity. Faced now with limits in your resources, the reduction of wealth in your nations and the ever-growing problem of human deprivation, individual freedom will be greatly moderated out of necessity. Humanity will have to have a greater social conscience and responsibility. This is part of your evolution as a race, and it is a very important one regarding the kind of future you will create for yourselves and the nature of your engagement with the Greater Community.

That is why the allies of humanity are seeking to encourage the preservation of freedom and the awareness of a deeper Knowledge within the human family. They want to have other free nations in their region of space, not because they are in love with humanity as much as for their own stability and security. They have a natural interest in promoting freedom in the universe, and to have another free society in their general vicinity is of great importance to them. They also see the value and the potential for humanity to evolve into a very magnificent race where individual capabilities and the benefit to society have great advantages. But it will not be easy to accomplish this.

That is why the allies of humanity have sent a set of Briefings to the world to help prepare humanity for the risks and dangers of encountering the Greater Community at large and to emphasize the development of individual talent and ability and the cultivation of Knowledge—a deeper spiritual Knowledge—that are at the core of any free race's existence. Humanity has all the great qualities that give it a chance to evolve into a free and self-determined race functioning in the Greater Community. But the requirements are tremendous, and education is of the greatest importance.

Given the difficulties of facing the Great Waves of change—environmental deterioration, the decline of your resources and the resulting problems in food production, the distribution of essential resources for people, the risk of pandemic illness and the threat of war and conflict—all of these things represent the immense danger that humanity now faces and will face increasingly as you proceed. And the danger of foreign intervention, by economic Collectives and other groups who seek to gain advantage and persuasion here for their own interests, adds to the complexity of the situation that humanity is now facing.

Some people believe that the free nations should come and

defend the world. But the free nations, to remain discreet, would not do this. They do not have the military power to do this either. It is humanity itself that must earn the right of freedom in the universe. It is the will, intention and commitment of the human family that will establish you as a free race. If you had to live under the protection of a foreign power, that foreign power would have to control human awareness and human governments to a very great degree. No nation that is free is equipped to do this, nor ultimately is it in your best interests.

You must grow up eventually. You must become mature, powerful and discreet. This is the challenge and difficulty of emerging into a Greater Community of intelligent life. This is the challenge that humanity now has before it—a challenge with dangers, but a challenge with the greatest of opportunities.

Power in the universe is power
in the mental environment.

CHAPTER 7

COMPETITION, INFLUENCE AND THE MENTAL ENVIRONMENT

As nations evolve and achieve stability and security, it is recognized that the use of force is not productive in engaging with other nations. It is destructive, and it sows the seeds for future conflict. Because resources are so extremely valuable to an advanced society, the use of force is seen as only a last resort and an undesirable one at that. The emphasis then shifts to the power of persuasion and the importance of persuasion in preserving resources and valuable infrastructure. It is an emphasis that nations within your own world are only now beginning to discover.

War is destructive. It is destructive to both the winner and the loser, and it sows the seeds for future conflict. The damage is not only to material things, but to the nature and character of entire nations of people who are traumatized. Humanity has not matured enough nor has grown in wisdom enough to recognize the absolute truth of this.

That is why outright warfare in the Greater Community, particularly on any kind of large scale, is extremely rare, particularly amongst secure and advanced nations. Their emphasis is on the power of persuasion. Because competition is so great in the Greater Community—the competition for resources, the competition for influence, the competition for trade and the competition for political power and persuasion—these are of paramount importance to any advanced nation, whether they be free or not.

It is because of the emphasis on stability and security that this is the case. Competition, therefore, is immense, and the powers of persuasion and the importance of these powers become of primary importance for nations that are free and for nations that are not free. This represents an evolution of power and skill in the universe. While humanity is still exerting force and destroying its own infrastructure and resources in the process, advanced nations use much more sophisticated means. Time has proven this to be necessary.

Nations that are fundamentally warlike and destructive do not last very long in the Greater Community, for any nation to exert itself in this way in a highly populated region of space such as your own would meet a united resistance. With only very few exceptions in the universe can this resistance be overcome. What would one nation do faced with a thousand nations opposing it? And what could one nation do if its access to trade were denied? To become strong enough to overtake other nations, you must have resources from beyond your own world. You must have a vast network of resources and resource acquisition.

In an undeveloped part of the universe, where there are very few advanced nations and where trade has not been established and secured over time, it is possible for one nation to become powerful in this way. But eventually it will meet opposition, collective opposition. An empire might exist in an uncharted region, dominating many star systems, but if there is no one within its reach to overtake, it will eventually be limited by its own resources and by its logistical isolation.

This is simply a fact of life. It is growing mature. It is outgrowing the passions and aggressions of adolescence to a more mature and stable state. But in this more mature and stable state, competition is intense, and therefore the emphasis on the powers of perception and persuasion becomes of paramount importance. This now becomes how you contend with competing nations. This is how you contend with problems and difficulties.

This power of persuasion and perception then becomes the focal point. Knowing what others are doing, perceiving their activities, trying to discern their secrets and their technological developments, discerning their intentions, discerning their communications and discerning their diplomacy—all become a great emphasis now.

For this reason, many nations have cultivated a special class of citizens, individuals called "seers." These individuals are not developed spiritually, but they have great skill in the mental environment. Their role is to see, to discern and to interpret. In normal trade negotiations, what is written and what is said have to be interpreted very clearly. The long-range implications have to be discerned. The advantages and disadvantages have to be clarified. It is a matter of great importance. While outright lying and cheating on a large scale may be rare, subtle manipulations are ever present.

Nations not only want to discern one another's intentions and capabilities, but also to discern one another's secrets, unspoken agendas and long-range plans. You living in the world can understand this. This is what the nations in your world do regarding one another. Even nations that have very friendly relations are always looking and watching to see what the others will do and to discern the others' strengths and intentions. Amongst nations that are in opposition to each other in your world, this is certainly the case. The difference in the Greater Community is at the level of skill, and the powers of discernment and persuasion are much greater. The nations in your world have not recognized the importance of establishing seers.

Among seers, there are visionaries and there are locators. Locators are a very special class of seers that focuses entirely upon locating facilities and pinpointing specific areas and activities within a nation. This has been developed to a very high degree to fight crime, for example, to deal with insurrections and to deal with the problem of illegal drugs, which is a real problem in many nations.

Locators are very important. If another nation has a secret facility, and it is recognized that such a facility either actually exists or may probably exist, locators are brought immediately to try to discern its precise location. Because visitation to other worlds is highly restricted, except between nations that have had longstanding and compatible relationships with one another, the function of locators becomes ever more significant. It is a specialized talent.

Seers are used to witness negotiations, to review contracts, to serve as witnesses to diplomatic efforts and to sit in on councils. They are certainly used in judicial matters to discern the truth and the falsity of evidence and declarations. Even advanced nations that are not free recognize the limits of technology. Over time, it has become recognized that power in the mental environment—the power to influence thought and to discern thought—extends beyond the range of technological capability and enters a new panorama of power and influence.

Nations employ seers on their defensive perimeters to watch for any possible threats on the horizon or any possible change that could be a challenge or a difficulty for their nation. They are not analysts in the way that you would consider analysts—using their intellect, using technology and using patterns of thought to discern with accuracy truth that must be recognized. A seer is using a different kind of power and potency that an analyst could never use.

There are seers who specialize as interpreters. They will travel with diplomatic missions, serving to interpret not the language of another race, but its intention, its honesty, the truth of its statements, its strengths, its weaknesses, its anxieties, its concerns and its insecurities. This is extremely important in establishing trust with races unlike your own.

Even nations that have longstanding partnerships that have developed trust over time are still always watching one another, not because they have an inherent distrust of one another, but for the effects of influence of other nations upon their trading partners. Is their trading partner being seduced or undermined by another foreign power? Does that trading nation have social harmony, or are there dissident movements within this nation, and what would that mean? While nations exert control and many are very uniform, dissension and opposition still exist and revolutions do occur.

Humanity has only evolved to deal with itself. It has never had to deal with other forms of intelligent life, particularly other forms of intelligent life that carry great power and ability—not only technologically, but in the mental environment. In this respect, humanity is still a primitive race. It is only beginning to recognize the potency of power in the mental environment. It is only beginning to recognize both the constructive and destructive potential of technology. Yet the great frontier in the mental environment is only beginning to be discovered and valued amongst the leaders of human nations.

Because you are unskilled in dealing with foreign powers in the Greater Community, you have not yet cultivated the necessary discernment that is required to discern the nature and intentions of a race that not only looks different but is different—that thinks differently, that has different values, that has different priorities, that has different traditions and that has a different social structure, a different history and a different well of experience. Humanity has not yet learned that power in the universe is power in the mental environment. It still thinks of power in the universe as empires conquering and destroying one another. This represents a child's view of the universe.

Secrecy, deception, discernment, cleverness and persuasion—this is where nations can overcome or gain preeminence over one another. Exploiting the weakness of internal divisions of another nation; exploiting another nation's mythology, fantasies and religion; discerning another nation's intentions, secrets, abilities and disabilities—these represent power in terms of nations influencing and gaining preeminence over one another.

Many nations employ seers, and seers have very specific functions as interpreters, as locators, in serving defensive purposes, in serving business contracts and in overseeing a nation's own internal needs and potential for disorder. The free nations have seers as well. Their seers are employed for the same purposes, but they are guided by a deeper Knowledge, which can make them more powerful and more penetrating than the seers employed in nations that are not free. This gives the free nations power and a certain degree of advantage that must remain hidden from foreign awareness and scrutiny.

Ultimately, the most powerful races in the universe are entirely hidden. And if they have any trade at all, it is maintained with the greatest secrecy and usually carried out by other nations that support them, that act as proxies for them. If you attain great power in the mental environment, your powers would be sought by other nations. Your skills would be sought by other nations, and other nations will attempt to seduce you or induce you through promises of wealth, splendor or high social positions—whatever they can do to seduce you in order to gain control of these abilities.

The greatest expressions of power in the mental environment are guided by Knowledge, which is entirely ethical and entirely peaceful in intent. Individuals possessing this cannot be turned and cannot be seduced; yet they still must remain hidden. This is one of the great dilemmas of life in the universe, and it is true even within your own world, that those who are most powerful, those that have the greatest degree of skill, must remain hidden, or they will simply be used by political powers, commercial powers and the powers of religious institutions. They will be used as tools to carry out activities that are unethical and destructive. It is like the person who possesses great material wealth—very difficult it is to hide this from others. Even if you create the appearance of poverty, the evidence of your material wealth will always be there.

Therefore, if a nation possesses tremendous material wealth and resource wealth, and it seeks to remain free and beyond foreign intervention and persuasion, it must keep these things in the greatest of secrecy and not demonstrate them to any degree at all. This of course is nearly impossible when surrounded by trading nations or if you are involved in the activities of trade where everyone is always looking for wealth and advantage. That is again why free societies remain discreet and distinct and involve themselves in trade only to a very minimum, if at all.

This is more difficult for a nation living outside a highly inhabited part of space, where there is social order, to maintain its hidden position because others can intervene without restraint. Invasion occurs. But living in the shadows in the midst of highly developed parts of space where there is tremendous social order requires restraint.

If outside invasion is forbidden, then a nation can, according to the rules of engagement such as exist in your region of space, maintain a private existence with a minimum of foreign scrutiny. But even here, the seers amongst technological societies will try to discern the power and capabilities of the free societies. Their scrutiny must then be offset by the seers of the free societies. This is competition at another level. It is persuasion at another level. It is the fundamental problem of possessing knowledge and wealth in the universe.

How does one possess knowledge and wealth and yet remain free? It is a problem for which there is no easy solution. It is a dilemma in life. Not all dilemmas have solutions. Not all problems have solutions. Because humanity does not yet experience itself living in a competitive environment beyond the borders of its world, it has not had to evolve to deal with these kinds of problems. But eventually it will. You are having to do this even at this moment, since the world is undergoing intervention from economic Collectives, whose ethics you could not agree with and whose methods are entirely self-serving.

This is the beginning of growing up and dealing with life at a more mature level. Eventually, every race in the universe has to face this. It is part of your destiny.

The importance of persuasion is obvious to you now, but in dealing with other races whose temperament and intellectual abilities and orientation are far different from your own, the power of persuasion becomes much more complex and difficult. Persuasion within your own race is one thing. But between races that are dissimilar, it is entirely another matter. Your logic and reason may not work at all with them. Your ability to rationalize with them may be extremely limited. What they value and what you value may be entirely different. They may have a particular tradition or persuasion or set of anxieties that are very unique to their history and circumstances. To understand this, to comprehend this, to speak to this and to be persuasive here requires a level of educational skill far beyond what human diplomacy has ever had to establish.

If you could, imagine trying to negotiate with an intelligence that was like a dolphin, except that intelligence had technology and represented a significant power. You want to negotiate for trade, you want to negotiate to establish rules of trade or engagements or mutual needs for security. How would you do it? And how could you ever persuade that intelligence? Where is it weak enough to be persuaded? Where is its weakness? Where is its strength? This is why it takes such great skill to negotiate and communicate.

When you have larger assemblies of nations on councils—on trading councils dealing with problems in international trade or relations, problems of crime, problems of trafficking dangerous materials or drugs or illegal trade—it becomes very complex. You are now speaking to maybe fifty different nations using common trade languages which have been established over time. They are there with their interpreters. They are there with their seers. Everyone is trying to communicate together. Some use language. Some do not use language. They all use written language or symbols. They all understand an established trade language, whatever that language may be in a larger district. How do you come to agreement and consensus? This makes negotiations very long and complex. Yet sophistication and time have shown ways to accomplish things, even given these difficulties.

There is communication, and then there is persuasion. Persuasion requires a very refined set of skills. You want another nation to see your point of view. You want another nation to value what you value. You want another nation to come to terms that are agreeable and preferred by you. You want another nation to do certain things or not to do certain things. You want to forge agreements and councils.

This requires persuasion. This persuasion is not only based upon the strengths of one's argument; it is based upon the awareness of another's comprehension and skills, another's nature and orientation, another's strengths and another's weaknesses. Persuasion is carried out for entirely beneficial purposes, and it is also carried out for extremely selfish purposes, for the interests of one nation alone.

That is why in highly inhabited regions such as your own, where trade has been long established, innovation comes very slowly. Once a practice or a system has been established and maintained, it is very hard to change it. If it has proven itself to be stable and beneficial to the majority of participants, it is very hard to change it, and innovations or improvements that are recommended or introduced can be greatly resisted. Even necessary change, even beneficial change, can be greatly resisted. The more nations and individuals involved, the more difficult it is to create consensus. Only if a network of nations or large trading associations are threatened clearly by some external force, unless this occurs, it is very difficult to create change and even improvements in the methods and rules of trade and conduct and so forth.

Here you are dealing with nations that have very different social structures. Some of these structures would seem absolutely abhorrent to you. How do you communicate to other beings in this situation? Are you going to try to change them? Do you want to try to improve them? Do you want to recommend different alternatives to them? Are they open to this? Are they willing to consider it, or do they consider your recommendations to be an insult, an affront? How do you maintain trust? How do you present the possibilities for change in such a way that other nations are not threatened? The finesse of diplomacy here must be very great.

What do you do when you have a nation in your midst that is behaving inappropriately, that is behaving destructively, that is casting an influence out into the local Greater Community that is unwanted or that is perceived as a threat? How do you regard a nation that is undergoing internal turmoil or even revolution? Do you get involved? Do you take sides? Do you make any recommendations at all? Do you value the rebellious side? Do you think the revolution is valuable?

What do you do with a nation that is extremely oppressive to its own citizens in a way that is abhorrent to you? Do you maintain relations? Do you maintain trade? Do you try to exert influence?

What if another nation has wealth or resources that you desperately need? How will you persuade them to engage in commerce with you and to value what you have to offer in return?

From these very few examples, and there are certainly many more, you can begin to see the complexities and difficulties involved here and why negotiations can be slow and protracted and why the attempt to change the thinking or the operations of nations engaged in commerce or in mutual defense can become very difficult.

That is why a great deal of energy is exerted on behalf of diplomatic relations and why seers are employed. Sometimes they are employed publicly. However, sometimes the presence of a seer is seen as an act of distrust. In many large chambers where decisions are made, seers are not allowed, for it is seen as an act of distrust, the evidence of distrust, to have a seer with you.

The injustices of a particular nation, the oppressive nature of certain nations, are tolerated under these terms because these nations are needed for trade and security. It is a very imperfect system, and it has many unfortunate oversights.

This is another reason why free nations try to stay out of these larger, economic involvements. Yet even for them, diplomatic relations are of the highest importance. The employment of seers is of the highest importance. The employment of skilled interpreters is of the highest importance—to secure one's position; to establish an autonomous region for yourself; to prevent intrusion; to provide overall security, not only for yourself, but for your entire district; to deal with biological contamination, which is a very serious matter; to engage in essential trade if necessary; to acquire the latest medications for your people and technology for producing food. All this is very essential.

Even free nations, should they be highly successful in achieving autonomy, still have needs that they themselves cannot meet. How should trade be conducted then? Who is a trustworthy trading partner? How will you engage in trade to keep others out of your environment? If your lifestyle is entirely different, if you are living in a much freer way than anyone else around you, how do you deal with their presence in your environment? How do you present a beneficial image? How do you present yourself as being non-threatening to nations that are not free?

Because of these difficulties, power and emphasis in the mental environment become significant. The mental environment is the environment of thought and influence. Into this environment, not only do you direct your thinking toward certain individuals, you also create thought forms. Thought forms are potent organized thoughts that can have an influence on the thinking and the emotions of others. They are not real physical objects, but they act in a way that is similar to an object.

For instance, in the mental environment, to stop the penetration of a seer, you must create a shield; you must obstruct a seer's intrusion with a counterforce. Seers between different nations can literally be in battle with one another to counteract each other's movements and discernment. If another nation thinks you have a secret facility that possesses power or potential that they are interested in, you must create a counter thought that there is no such thing. If they try to locate something that actually exists, you must persuade them that it is somewhere else.

It is an engagement on another level. It is part of the burden and responsibility of a seer not only to see, but to defend against the intrusions of others in the mental environment. The seer must also be able to project images and associations that you want another race who is scrutinizing you to have, not only to keep your secrets, but to give an accurate and preferred image that you want others to have of you.

Because travel between nations is so restricted, the ability to see into other nations becomes a very great emphasis. Since you cannot visit and go anywhere you want in someone else's world, the ability to see into that world becomes very important. Likewise, their attempt to obstruct your viewing of their world becomes very important for them. This is particularly an emphasis for nations that want to remain self-determined. If you are merely a client state of another power, you have no secrets, but you also have no choices, for your range of self-determination will be extremely limited.

It is a problem that requires adaptation and the cultivation of skills. Growing from being an adolescent person to being a mature adult requires discernment, discretion and the development of skills if one is to be successful in a mature environment. If you act in an adolescent manner, others will not respect you, others will take advantage of you, and you will suffer many consequences and will have very little social power.

Within all of this is the introduction of powers and races from beyond one's district that are providing opportunities for trade. How do you regard them? Who are they? If their home world is beyond the normal reach of your travel, how do you find out about them? Should you include them in your network or should you not include them? Are they truly beneficial or are they not beneficial? Do they pose a risk of biological contamination, and, if so, how can that contamination be counteracted? What are their strengths? What are their weaknesses? What are their intentions? If they represent an entire other district of organized states, what is the organization of that? How do you relate to that? What are the influences there? What are the powers that exist there? What is their strength in the mental environment and so forth? You see, this becomes very complex.

If there are disparities in technology, then the gap must be closed as quickly as possible. That is why within a region, there is usually a technological parity, in spite of the fact that nations have secrets—secret skills and secret abilities that they do not want to be shared, copied and used against them.

Technology can be purchased, it can be copied, it can be stolen, and it is acquired through all these means. That is why once one nation in a region develops new technology, it is not long before everyone has it. To keep it secret is very difficult because if you keep it secret, you cannot really use it. And if you use it, soon it becomes known to others and loses its potential advantage in negotiations or relations with other states.

This is why no nation has absolute power because it could not sustain itself over time. It would fracture from internal divisions, internal rebellions and from problems in resource acquisition. Its skills and powers would be discerned and copied and its technology stolen. To reach stability and security, most nations will share or sell their technology if they cultivate something that others do not have. It is very difficult to keep it secret, and if you try to do so, then it arouses suspicion and distrust, inquiry and discernment.

Along with this, there is also a fascination with the internal workings of other nations—their social structures, their social problems and the difficulties that their leadership undergoes. Rumors are created. There is much speculation. Nations take great interest in the peculiarities of other nations, in their social problems and in their corruption. This is done to promote one nation's sense of unity and superiority. It is also done because intelligent beings are by nature curious. And because intelligence breeds curiosity, that is one of the problems of collective breeding. How do you keep a bred segment of the population from becoming curious, interested, stimulated and tantalized by things they might see around them?

Intelligence breeds curiosity. It also exposes one to greater degrees of influence and stimulation. Nations become very interested in one another's peculiarities, difficulties, corruptions, problems, conflicts and so forth—not only from a political standpoint, not only in the halls of government, but amongst citizens as well, who love to indulge themselves in the problems of others or in the suspicion of problems of others.

Therefore, there are many rumors, of course. And much of this is published for public consumption. Large nations, if they have independent citizens, have large media networks that are highly controlled. But, nonetheless, rumors and even real secrets emerge and are hard to keep out of public scrutiny. In freer nations, there are very few secrets that are kept from the people to limit speculation and to establish unity around wisdom and shared concern for security and for well-being. It is an entirely different emphasis than is exerted in states that are not free.

Humanity is emerging into this environment of persuasion and power in the mental environment. That is why you must be extremely discreet. That is why you cannot be broadcasting your entire society out into space through your transmissions. That is why eventually you will have to have seers who are skilled, locators who are skilled and interpreters who are skilled—both to deal with internal affairs and to deal with the immense challenges of functioning in the Greater Community.

Freedom is not a right
in the universe.
It is a privilege.

CHAPTER 8

FREEDOM IN THE GREATER COMMUNITY

Freedom is rare in the Greater Community—individual freedom and the existence of free societies which value the potential and the creativity of the individual—particularly as it pertains to the welfare of a nation and a world.

You might wonder why this is so, thinking that the development of freedom really represents the goal of evolution in a higher sense. But freedom is rare, freedom in any society, even in a primitive society. It is rare because of the difficulties of living in the physical environment—the challenges of survival, the acquisition of resources, competition with others, the threat of war and annihilation, the problems of governance, the size of civilizations and the rise of technology. All these things, to whatever extent that they exist in any world or in any group, tend to have a limiting factor on the value and the recognition of the importance of individual freedom.

Groups must work together to survive. And surviving in the Greater Community, even as a technologically advanced race, is not an easy proposition. If you become dependent upon other nations for the very resources you need to live and to function, then your existence is always jeopardized by any interruption in trade and commerce or the threat of exclusion that can be placed upon you by other nations. To lose your self-sufficiency is to move into a position of greater insecurity.

People in the world today associate advancing technology with advancing security and often for very good reasons, for many people in the world today have an assured supply of food and can gain access to resources that only the nobility in times past could acquire. A great deal of wealth and a great deal of stability have been created amongst nations that have this kind of affluence. But should the resources that support this affluence begin to decline, as is happening in the world today, then you can see how easily your wealth, your security and your stability can begin to diminish as well.

Freedom is not a right in the universe. It is a privilege and a luxury. Your values may argue with this, but your expectations eventually must conform to the real circumstances of life. Individual freedom is extremely valuable, but it is not guaranteed, and you cannot claim it as a right. You will be shocked to find, as you begin to learn about and even experience the Greater Community in which you live, how rare freedom really is.

Freedom itself is always relative. You never have complete freedom of movement and expression. In concert with others, you cannot do whatever you want or say whatever you want, and you understand this. You will never have this freedom if you are living in association with others and if you are functioning to survive as a group, as a nation and as a people. Therefore, freedom is always relative to your circumstances, to your affluence and to the degree of stability and security you have been able to establish and to maintain over time.

Freedom for the individual becomes a problem in terms of stability. Nations that are more democratic tend to be internally more dynamic and more creative, but in a sense more unstable as well. If individuals have the freedom to express themselves, they also have the freedom to exert social power and to change the structure of their society. This is beneficial if the society has an adequate provision of resources and can sustain this provision over time. But as humanity will soon discover, as these provisions begin to diminish, the circumstances and the allowance of freedom will diminish as well.

Nations that are under great pressure to provide essential resources generally do not allow a great deal of expression of personal freedom. They cannot afford it. They cannot afford the social disorder this creates. They cannot afford dissension. They cannot afford to have their people opposing one another—internal strife and discord, factions working against each other and special interest groups overriding the needs, the rights and the concerns of the general population. This all creates instability.

That is why the freedom that must be emphasized is a greater internal freedom—the freedom to find the way to Knowledge, the deeper intelligence that God has placed within you and within all sentient life. For your circumstances may require great limitations in what you are able to say, to do and to express in your society. Even in the world as it undergoes the Great Waves of change, this will be the case. The larger a society becomes, the more restraint there will be upon personal freedom. The more a society must stabilize itself for its own survival, the more restraint there will be on personal freedom.

In a world such as yours, with an ever-growing population and a shrinking well of resources, you can begin to see how personal freedom will be limited in the future, circumstantially limited, limited by necessity. You will not have the wealth or the social power to do the things perhaps you could do before if you were living in an affluent nation. People do not make this association. They do not see that plundering their world's resources is actually creating the very set of circumstances that will rob them of their personal freedoms and advantages.

Nations that have achieved a more mature, stable state also have limits on freedom. But in free societies, the power of creativity for the individual is still highly regarded. Yet the freedom for people to be reckless, destructive and chaotic is most definitely suppressed. The stability and security of the nation now become essential, even in a free society. So while you may have the freedom to express yourself and to contribute your unique gifts for the welfare and the benefit of everyone, you do not have the freedom to be reckless, destructive or chaotic.

Perhaps you will view this with anxiety, but you must see the necessity of this. As humanity begins to face the fact that it has grown beyond the limits of its resources, you will see the power of the needs of civilization itself—the needs of the human family—to prevail over the rights and privileges of its citizens out of sheer necessity and that the expression of human freedom must become founded on a deeper foundation of Knowledge within yourself.

Humanity has a great benefit here and a great opportunity to maintain the value of individual freedom and creativity in order to generate the necessary advancement in all aspects of your life. But, circumstantially, you will lose many privileges in the future. At this moment, you can use your personal conveyance to go wherever you want, in most places. In the future, you will not have a personal conveyance. You simply will not be able to have this kind of flexibility, not because someone is oppressing you, but because the resources do not provide for it. The circumstances do not provide for it.

You have created this condition. You and everyone else created this condition, this limitation. Because human civilization will be under great stress and great peril, extreme measures will be taken to limit internal conflict and dissension. In many cases, this will be unfortunate. In many cases, this will be anti-progressive. In many cases, this will be detrimental. But you cannot have disunity in the face of adversity, and great adversity you will be facing.

Advanced nations in the Greater Community face adversity as well. In most cases, they have outstripped their world's natural resources. They have developed technology that depends upon foreign trade and foreign manufacture. They have lost the freedom to sustain themselves, to be self-sufficient. Now they must meet the terms and agreements placed upon them to gain access to the resources they need from abroad. Now they must engage in the complex and taxing endeavor of negotiations, persuasion and power in the mental environment.

You can see here where power and affluence lead to the loss of personal freedom. It begins to work in the other direction. Most nations that have become technologically based have lost their essential freedom and integrity because they have overwhelmed their home world's ability to provide for them. They also have been seduced to acquire technologies that they themselves could not produce, using resources that they themselves could not produce. As a result, they have become dependent upon foreign powers and forced into engagement in the complexity of trade in the Greater Community—subject now to councils and agreements established with other nations, subject now to the rules of engagement imposed upon them by others. If a world becomes overpopulated and outstrips its resources, it will either fail and decline or come under the persuasion of foreign powers. This is the inevitability of following a path of this nature.

In your world, amongst the wealthy, freedom is regarded as a right and is taken for granted. But you do not see that it is all based upon the acquisition of resources, upon the wealth of resources, upon the wealth of your world. The fact that you simply cannot go out into the universe and gain everything that you have depleted here on Earth is a reality of life that has not yet been recognized. If this were recognized, it may change human behavior and human expectations very significantly, and that is the power and potency of the teachings that are being presented here.

You do not want to drive yourself into a position of deprivation, or you will have no power or efficacy in the Greater Community. You will have to agree to whatever terms are offered you to provide you with the very things you need to live, the very things that you have exhausted here in the world. Here your environmentalism is not simply based upon aesthetics or spirituality. It is based upon the vital necessities of life.

Free nations in the universe value the individual and take advantage of the individual's talents and natural inclinations. That is the great difference between a free nation and a nation that is not free. In all cases, however, in higher social orders, dissension, conflict and an individual's destructive tendencies are limited and, in many cases, greatly suppressed. The great difference is the valuing of the individual's innate capabilities and supporting the individual to participate in society based upon the contribution of those innate abilities.

In a nation that is not free, you are simply assigned a role based upon your social standing—the standing of your family, the standing of your parents, the standing of what your society requires from you. This has nothing to do with your individual talents and abilities. The only exception made here, in a nation that is not free, is if you have or demonstrate clairvoyance, in which case you may be prepared to function in the diplomatic corps, which is an entirely different education. Nonetheless, you are not free to choose or determine your destiny in a nation that is not free. It is determined for you entirely. And the determination is based upon your social standing, not upon your individual talents or inclinations.

In the future, should humanity navigate the Great Waves of change, you will be living in a much more structured world. You will be living in a world that will have undergone tremendous travail and deprivation. Nations will have to unite with each other to maintain the well-being of humanity. This will be a very different world from the world you experience today. You will have very little social power. You will have very little wealth. You will have very little freedom of movement, either because it is denied you or simply because you do not have the resources to enjoy it or to utilize it. Life will be far less free than what you may experience today in an affluent society. In many parts of the world, you would be assigned to work wherever you are needed, without regard to your own talents and inclinations, unless you had a special set of skills.

People will have very little wealth, except for a very few. There will be very little consumption compared to what you experience today in your wealthy nations. You will be living with a memory of the past, of better times, remembering times of greater affluence and greater freedom, greater enjoyment— times that were not overshadowed by the great problems of the world and the degree to which they will occur in the future. You cannot escape this future, just like you cannot escape the reality of the Greater Community.

You are following a pathway that most nations have followed— gaining technological abilities, their growing populations outstripping their world's natural resources, being forced into deprivation, into greater control and being required to unite or to fail altogether. You will look back upon this past century, and particularly the latter parts of the twentieth century, with great longing. This will be true for so many people. You have squandered the wealth of this world. You did not preserve it, you did not sustain it, and now you must deal with the consequences. Part of the consequence is the tremendous loss of freedom. You do not see this yet because you do not have the wisdom to see it. It has not been demonstrated.

There are people who will say, "Well, anything would be better in the universe than what we have here," but these people do not recognize what a completely foolish statement this is. To live in a world that is not free would be unimaginably difficult for you. And depending on the severity of this, your life would be controlled. Should you show any aberrancy, you would either be imprisoned or destroyed. You would have no right to protest. You may have very little opportunity to improve your circumstances or even to make a recommendation to improve the circumstances of everyone. You would have no personal choices. You would be assigned a role. You would be assigned work. You might even be assigned a partner. It may be prohibited for you to generate offspring if your genetic qualities were not valued. To have a child you would have to gain a license from the state. You would have to have an approved partner. Your children would be taken from you at a very early age and educated according to the needs of the state. Such things are very common amongst technologically advanced races.

You can begin to see here what an awful prospect this really is. Yet this is common amongst large technological societies. Even in free societies, individual aberrancy and destructiveness are extremely limited. You cannot be reckless and chaotic if that threatens the welfare and the stability of your nation.

You can begin to see here that the requirements of living in the Greater Community, even if you could evolve and sustain yourself as a free race, would be very different from what you experience today. There would be greater social welfare, but you would not see the personal wealth that you see today. You would not have the personal freedom that you might enjoy today. There would be sufficient food and medical provisions, but most individuals would not have much personal wealth. Your freedom of choice would be really limited. If you had a unique talent that was recognized, you might be offered a very limited possibility of employment. This is true even in the freer nations of the universe.

You can see here that life in the universe is not based on human values and that the emphasis on necessity is tremendous. Even in a society that has high values and ethics, necessity requires a certain amount of conformity to the needs of everyone. Free nations must be united to survive, to offset foreign influence, to maintain stability and to maintain their security. How unlike this is from the world you see today—where nations are competing with one another, where there is immense suffering and poverty, where there are very small groups of wealthy people who indulge themselves endlessly in their own pleasures and obsessions.

This would be very rare in an advanced nation in the universe. You may have a ruling elite, and they may have privileges and wealth that no one else has, but in a free society, the discrepancy would not be as great as what you see today, nor would the abuse of this wealth be as great as what you see today. So you must adjust your expectations, for humanity is moving into a position where it will have to choose between individual freedom and stability and security. They will tend to offset one another. And you will not have the wealth to indulge yourself to the degree to which you have.

This is the world you have created. This is the path of evolution you have chosen, and you will have to adapt to the consequences. The adaptation will be very difficult, but it could be beneficial to you, for it will prepare you for the realities of life in the Greater Community, where a world such as yours—stewarded by a weak and divided race, however talented—would not survive for long in the presence of greater forces. Indeed, humanity has reached a point where greater forces will begin to exert themselves here.

Previously, for foreign races to gain biological resources from the world without creating establishments here has been very difficult to do, given the biological diversity of this world and the fact that most advanced races live in either completely or relatively sterile environments. As a result, humanity has been allowed to evolve with very little interference. But interference now is beginning and will be continuous from here on out.

It is a very difficult transition you are beginning to undergo, one that most races will have to face. This transition will require human unity and cooperation. It will not tolerate unwanted dissension to a very great degree. It will limit human wealth and freedom. And it will place you under increasing pressure and influence from the Greater Community itself.

You can begin to see here that there is nowhere for you to be reckless, there is nowhere for you to run and hide, there is nowhere for you to live in fantasy with dreams of personal acquisition. You are not going to go out and plunder the Greater Community. The Greater Community is not full of ignorant, isolated, primitive tribes of beings. To the contrary, it is a mature environment—an environment that has gained great stability, an environment that has required great restraint, a competitive environment, an environment of influence, an environment of discernment—an environment that humanity will have to contend with increasingly from this time forward.

When We begin to speak about the spirituality of life in the universe, We will speak of freedom in a more complete sense—the power and presence of Knowledge within you, the greater intelligence that God has placed within you to guide you, to protect you and to lead you to great accomplishments. This is a different kind of freedom than social freedom and can exist in environments where social freedom has been greatly limited. It is this greater freedom now that you must embrace and cultivate, for it will give you advantage and precedence in a world in decline. And, as you will see, this greater freedom will be immensely important for humanity in determining what kind of future it will have within a Greater Community of intelligent life.

PART TWO

THE SPIRITUALITY OF INTELLIGENT LIFE IN THE UNIVERSE

Do not worship the form of religion.
Seek its essence, its power and its intelligence.
This is the highpoint of religion in the universe.

CHAPTER 9

RELIGION IN THE GREATER COMMUNITY

Religion exists in countless forms in the universe, from very simple tribal practices to practices that are undertaken by large empires and by large assemblies of nations that have adopted a singular practice. While there is diversity in the universe, there is rarely diversity within one world. As nations become more technologically focused and as resources become more rare and difficult to obtain, uniformity and singularity become the central emphasis, and here religious institutions almost inevitably end up being tied to the state. Some nations become entirely secular, never having had a strong religious tradition of their own. Or, in the case of being colonized by another nation, whatever religious tradition they had was assimilated and often erased from the public consciousness.

For nations that have maintained a religious tradition, it becomes wedded to the state so there is no competition, duality of power or crossing of intentions. Unfortunately, in this situation, religions end up losing their primary spiritual focus. They lose their essence and become vehicles of worshiping either the leadership or the race itself—praising the beneficence, the excellence and the superiority of one's race, the inevitability of one's success and one's rightful superiority over other nations. When religion is united with a political system to this degree, it rarely survives the process and ends up being a ceremony for government and another method to keep the people focused and oriented towards the wishes and dictates of the state.

Religion has ended up in this situation in so many places. Where it has been able to maintain a certain degree of autonomy from the political system, it has had to focus itself entirely upon ceremony and transcendent matters—never criticizing or creating doubts about the dictates of the political system, never raising ethical questions about its treatment of its own citizens or its views or treatments of other citizens. As long as religion in these circumstances remained fundamentally transcendental, it was not a problem.

Nations where there is a diversity of religious practices are extremely rare, and there is great resistance to the introduction of another nation's spirituality. That is generally resisted intensely amongst trading nations. The sharing of religious views and ideas rarely occurs and is kept out of the context of commerce.

While religion here has become extremely limited and controlled, there is a greater spiritual practice that is happening in many places in the universe—often in secret, often under clandestine circumstances. It is a practice of something much more fundamental. It is worthy of your attention and important for humanity to discover. For it is here—if humanity is ever to establish a deeper connection with another nation in the universe—through this deeper practice and awareness that a greater association may be established.

The Way of Knowledge is practiced in many worlds, Knowledge being the greater spiritual reality that the Creator of all life has imbued in intelligent life throughout the universe. Knowledge remains a potential in all intelligent life. Even in intelligent life that has been bred and genetically modified for specific purposes, it is still a potential, even in these circumstances. This potential, however, does not mean revelation. It does not mean awareness. It is but a potential.

There are certain small groups and free nations that have achieved very high states in this practice—a practice that is largely without ritual, that does not praise idols and that does not worship specific teachers, avatars or spiritual representatives. While these things may be added in certain cultures, according to the needs and traditions of those cultures, the fundamental emphasis is on the discovery and the expression of this deeper spiritual power and intelligence that is embedded within the individual.

This intelligence is extremely valuable in the Greater Community, particularly for free nations, because it is the only part of the individual that cannot be influenced or manipulated. It remains entirely honest and pure. Even if the person who possesses this Knowledge turns against it and denies it, or loses awareness of it completely, Knowledge within them remains pure and undefiled. It can only respond to God and to the presence of Knowledge in others.

This is why Knowledge is viewed as such a tremendous danger and a threat to oppressive regimes, to hierarchical societies, to those who engage in collective breeding and to nations that want to retain utter control over their citizens. If they are aware of Knowledge at all, it will generally be viewed as a threat, as a competitor for authority, as something that could seed dissension or arouse hostility within their populations. It could arouse the desire for freedom, arouse the desire for individual self-expression and arouse indignation at the long weight of oppression that has been placed upon them and its extremely damaging effects. This is why the practice in The Way of Knowledge is carried on in secret and in private. Only amongst free nations can it flourish and be seen as a benefit to the nation, instead of a source of competition to those in power.

Nations that are guided by Knowledge have special skills and powers, which they must protect from intervention and scrutiny from the outside. This can represent the secrets of a free nation, for Knowledge can give rise to great inventions, to great insights and to great powers of perception. Those who are seers with Knowledge can have great efficacy and great range and penetration in their perception. But this requires Knowledge as a foundation for one's thinking and behavior for such creations to emerge and for such skills to be developed.

Knowledge is entirely too revolutionary for any kind of a dictatorship. It has a resonating effect with others, and if it is demonstrated or expressed, it inspires and encourages others and makes others aware of their restraints and lack of freedom. This is why the practice in The Way of Knowledge has been suppressed and even crushed in many large and powerful nations. Those who practice this essential spiritual focus must do so in great secrecy, functioning in deep networks of practitioners and even communicating with other practitioners in other worlds whenever that is possible. Sacred texts are smuggled to other nations.

The power and the revolutionary nature of Knowledge make it extremely potent. Yet it cannot be misused by heads of state or by powerful commercial forces. You cannot use Knowledge. You can use psychic abilities, you can develop a seer without Knowledge, but you cannot use Knowledge. It is an intelligence all of its own. It cannot be used and manipulated to provide advantages or domination over others. This is what makes it holy and pure and authentic. Even amongst the practitioners, they will begin to see in their studies that there is a difference between the mind that they think with—the mind that has been conditioned by their culture and worldly experience—and the greater intelligence of Knowledge within them. To the unwise, Knowledge looks like a great power that can be used for many applications. But the wise understand that this cannot come about and that Knowledge is not the same as psychic ability or developing one's sensory skills.

For a free nation, the cultivation of Knowledge amongst its advanced practitioners becomes extremely valuable in discerning the intentions of others from the outside, in recognizing challenges or difficulties for the nation and for providing beneficial forms of education for their citizens. In a community of Knowledge, you want everyone to become strong with Knowledge because that increases the power of Knowledge tenfold. Now you have many eyes looking, many ears listening; you have many minds that are attuned to a greater set of abilities. This group mind in Knowledge is cultivated in certain rare and unusual societies with great power. It can discern, almost without evidence, the mysteries and the secrets of others. And it can be used to foster the most beneficial forms of scientific and philosophical development.

In nations that are not free, there is also a group mind process that has spread considerably amongst this part of the Greater Community, where groups of individuals will focus on trying to solve a problem or to penetrate another's reality. This has great efficacy if it is directed with skill, even if it is used for purposes that are unethical or destructive. Whereas the power of technology has its limits, the power of the mind has rarely been fully developed by anyone. The mind is seen here not as the source of power, but as the medium for power. That is a very important distinction.

There is a network of the wise, and there is a network of practitioners in The Way of Knowledge. Most have never met one another.

But the seeds of Knowledge were planted in their cultures and are supported by spiritual forces who exist beyond the visible range—who encourage the discovery, the development and the expression of Knowledge within many, many worlds. For a nation can direct the thinking and behavior of an individual, but it cannot overcome Knowledge within that individual. Once an individual can begin to sense that there is a greater presence and a greater intelligence within them and can begin to receive signs and guidance from this intelligence, they have begun the long process in regaining and developing the most fundamental freedom.

In general, Knowledge will have to be shared in secret. It is not for public consumption. Even in your world, there are very few people who are really ready to become a student of Knowledge and to learn The Way of Knowledge, The Greater Community Way of Knowledge. The requirements upon them will be great. They will have to learn to rise above their cultural conditioning, their attachments, their fears, their hostilities, their grievances and learn through experience and through many stages of development to experience the presence of Knowledge and to express its guidance, its power and its contribution in their life in service to others. Such a gift is being presented into your world now for the first time, a Greater Community Way of Knowledge, as it is practiced in the Greater Community. But it is very challenging. It is not that it is so difficult; it is that people are so dedicated to other things.

To be guided by Knowledge is to be in a very natural state. It is to experience one's own integrity. It is to have one's life and priorities in balance. It is to have one's relationships in balance. It is a very natural state. But people are living in very unnatural ways—being guided by their social conditioning, by their beliefs, by their ideas, by their grievances, by their political persuasions, by their nationalistic feelings, by their sexuality and by their hatred. It is shifting one's alliance, then, from any of these various things to the presence and power of Knowledge that represents a great transition.

There are those in the universe who have been practicing The Way of Knowledge for generations who have achieved a very high state of clairvoyance, who have been able to produce marvelous creations in technology and art and in materials and processes that are beneficial to their societies. But such contribution can only flourish in a free and respectful environment. Given that most nations evolve around their technological development and their need for resources, forcing them into a situation which requires extreme control and adherence for their populations, Knowledge has not been able to flourish in many places. But it remains the most fundamental spiritual practice in the universe. It is the source of all your world's religions. It is the intention that God used to initiate them to be Ways of Knowledge.

Even in your world, religious traditions have been aligned with political forces, have been used by ambitious individuals and have become institutions in and of themselves, seeking worldly power. The Way of Knowledge has become hidden, only accessible to a few who do not have these ambitions, who are free enough and who have the inspiration to search for the mystery of their existence.

Never assume that technological advancement leads to spiritual awareness, for in most cases it goes in the opposite direction. As evolving races begin to develop technology, or to receive technology from the outside, they begin to experience tremendous power over their environments. They are able to generate wealth and privileges, to gain access to new resources and to have new kinds of possessions.

This is the fixation of the mind, but it is not of interest to Knowledge within them, which becomes silent and latent within them, as other goals and other gods are sought.

The power and presence of Knowledge will be instrumental in guiding humanity towards maintaining its freedom and its self-sufficiency. It is the counterpoint to the greed, to the corruption, to the oppression and to the abuse that are so rampant throughout human societies and cultures. The quest for wealth, the quest for power, the quest for beauty, the quest for control—these are the obsessions that dominate the mind, but not only the mind of humanity. They are what dominate the minds of all beings living in the physical universe.

You at least have the freedom to think some of your own ideas in most of your nations, and your religious traditions have been kept alive, which holds promise for opening a doorway to Knowledge. You have a great benefit here, and yet even now your peoples are turning away from the mystery of their lives to seek greater power, greater wealth, greater leisure, greater pleasures—seeking to increase their possessions, seeking to increase their separation from others and their insulation from others. So you can see this emerging pattern even within humanity—fascinated with machines, fascinated with technology, believing technology will solve all the problems that technology creates.

People lose their sense of themselves, of their greater strength, becoming ever more reliant upon their machinery and their creations and technology to solve things for them. You can see the emergence of a new religion here, the religion of technology. But humanity is still fortunate in that the power and the presence of Knowledge have not been lost here and are still practiced by many and experienced by many.

You do not realize how precious this freedom is, in contrast to so many other nations in the Greater Community where freedom is either extremely limited or nonexistent. Ultimately, humanity may have a great gift to give to others, if it can survive the Great Waves of change and if it can maintain its integrity as it emerges into a Greater Community of intelligent life.

Knowledge is the most powerful force in the universe. But it is only known to a few, and it cannot be used and manipulated by the deceptive, the corrupt or the unwise. With it, you would be able to discern the nature and intentions of any who visit your world. You would be able to discern, in accordance with others, in alliance with others, all the solutions that humanity will need to bring to bear to meet its great challenge in facing the Great Waves of change.

With it, you will find the strength to persevere. You will not give up. You will not capitulate to offers of peace, prosperity and technology from opportunistic races in the universe. With it, you will maintain the awareness of humanity's essential value, and you will not trade this value for anything—not for power, not for technology, not for domination. You will moderate your behavior, even radically if necessary, to sustain the world as a habitable place for the human family.

You will not fall under the seductions and inducements of the many races who will seek to gain access to your resources, who will want you to become dependent on their technology—engaging in their networks of trade, offering you trinkets from space, offering you the promise of wealth, saying that they live in peace and they can give you peace, offering to help you lead your governments, offering to teach your people their religions, offering to take the reins of the world on your behalf. You will not be deceived by these inducements, these offers and these promises because Knowledge cannot be seduced and cannot be manipulated by any force, be it human or extraterrestrial.

For the experience and practice of spirituality to be shared in the universe, it must be translatable. To be translatable, it must be without adornments, without being encrusted with history and personalities, magic and miracles. It must be pure and efficacious. Otherwise, it cannot be translated from one world to another. It cannot revolve around worshiping heroes or heroines. It cannot revolve around the worship of an individual or graven images because it represents something pure and essential within the individual. It must be discovered through experience by passing through thresholds in one's life, through service to one's nation and service to life.

The advancement of science and technology alone will lead you away from the power and presence of Knowledge. Only if science and technology are in service to Knowledge, Knowledge being the prime motivation for the cultivation of science and technology, can they be truly beneficial. It is the motivation that matters. And this motivation either comes from the mind or from the deeper mind of Knowledge within you.

If you were to meet another being from another world—a being who did not share your language or your appearance, a being who did not breathe your atmosphere, a being whose history was far different from your own, whose environment was very different from your own—you could still communicate with this being through Knowledge. You would understand each other's intentions, and, if you were well cultivated, you could communicate to each other through the language of Knowledge, which is a language of impressions, images and signs.

Thus, Knowledge becomes the great peacemaker, the great equalizer, that which is common to all intelligent life, that which leads to the freedom of the mind and the freedom of one's life. Even if you are living in a constrained set of circumstances, which will increasingly be the case in your own world, even in your wealthy nations, with the power of Knowledge, you can still live a fulfilled and fulfilling life, with relationships of a very high caliber, in service to others. This is essential and fulfilling to you and necessary for them.

Therefore, you have religion with its complex ideologies, existing with or without a connection to Knowledge. You have religion as an appendage to the state in many worlds. And you have states without religion at all, the state being the religion, the intentions of the state being the focal point for its people and populations.

Knowledge is everywhere in the universe, but it has not been discovered or practiced by many. It represents humanity's greatest opportunity because it will take a long time for you to have technological parity or to develop the kind of social cohesion you will need to deal with the Greater Community, where social cohesion is the norm. But Knowledge will save you from error. It will prevent you from making critical and fatal errors. And it will prevent you from giving away your authority, your integrity and the reins of this world. It will give you hope when things appear hopeless. It will give you confidence when no one else around you has confidence. It will give you courage when others are terrified. It will enable you to remain free in an environment where others are not free.

This is religion in the Greater Community—religion that is translatable, that is universal and that is primary and essential. The customs, traditions and historical accounts that are added on to this are secondary and often of little value. Do not worship the form of religion. Seek its essence, its power and its intelligence. Knowledge is highly intelligent, far more intelligent than one's thinking mind.

This is the high point of religion in the universe. It is not an escalation in terms of theological understanding, or in terms of ceremony, or in terms of symbology, or in the construction of temples and complexes. Its elevation is its practice in an essential form, maintained in secret and in free worlds, taken to a high level of expression and experience in the presence of great teachers, practiced away from the scrutiny of government and commerce, protected from the ambitious and the corrupted, the self-seeking and the self-serving. This is religion in the universe.

Knowledge is the most powerful force
in the universe.

CHAPTER 10

THE POWER OF KNOWLEDGE

Developing skill in the mental environment is a great preoccupation amongst technically advanced nations, for it holds great promises of power and discernment and, in some cases, even the outside control of the experience of others. This power is something that humanity has only begun to discover and to develop. While influence is practiced in many ways in your world, certainly in every family and in every nation, it still represents a set of skills that humanity is only beginning to discover and to discern.

It is a power. Because it is a power, it requires great responsibility and the guidance of Knowledge to be used correctly and appropriately. Like any power, it can be used for good or ill. In the Greater Community, the power of the mental environment represents a real frontier beyond the limits of technology, for in regions where nations have much contact with each other, they have over time developed parity in terms of technology. It is rare that one will have a great technological advantage over another for very long.

So the subtle skills employed to discern the intentions of others, to plant ideas in the minds of others and to utilize one's knowledge of another's nature, skills and predispositions become of great importance in seeking advantage and in avoiding disadvantage. These skills are being brought to bear upon humanity at this time by those commercial forces who are in the world today trying to pacify humanity, trying to win human allegiance, trying to convince people in positions of power of the superiority of the Intervention's goals and the weakness of humanity's leadership and skills in dealing with the difficult times ahead.

These forces who are in your world are not military forces. They are depending entirely upon human acquiescence and thus bring their skill in the mental environment, which is the only real power that they have, to achieve their goals, to diminish resistance to their presence and to increase people's desire for their claims of leadership and authorship here in this world.

Here you can see clearly a manipulation of the mental environment. Yet you can see this in all of your commercial attempts at advertising and persuasion that are ever present around you, trying to convince you that you are inadequate, that you need this product or this service to look better, to feel better, to be better, to be happy, to be successful or to be spiritually fulfilled. All represent manipulation in the mental environment, tantalizing the mind to make you think you want something that you would not want ordinarily, to make you think you need something that you ordinarily would not need, or a need that you would fulfill in other ways.

The examples go on and on. They are endless in your experience. But in the Greater Community, the potency of this influence is greatly increased. Ideas are planted in your mind without the use of symbols. Others can manipulate your dream sequences. They can create reoccurring thoughts in your mind. If you do not know how to discern these from your own thoughts, then you will claim ownership for them and they will affect you. They will affect you, influence you and lead you to do things you would not do otherwise or to make decisions that were not in your best interests.

The power of inspiration, on the other hand, comes from Knowledge. It cannot be influenced. It is immune to this kind of persuasion. Even a Greater Community set of skills used against an individual strong with Knowledge would not be effective in changing their perception, their understanding or their experience. The influence would be felt, but it would be recognized as a perpetration from the outside.

This ability to be objective with one's own mind, thoughts and experiences is a very important part of practicing the Greater Community Way of Knowledge. It requires great discernment. It requires that you stand back from your thoughts to observe them rather than simply being governed by them or overtaken by them or tormented by them.

While humanity is now dazzled by its advances in technology, it has yet to learn the greater powers that exist in the mental environment, powers that will exceed your technological abilities. For in truth you do not want to become too technologically advanced, or you will lose your self-sufficiency. Whatever technology is adequate to provide for your peoples to provide stability, security and to maintain freedom in your world, would be adequate if you can create it and sustain it. But if you want things from beyond the world and can be convinced that you need them and that you must have them for your safety, for your security, for your advancement or for your enrichment, then you have stepped over a boundary, a boundary that Knowledge would discourage you from passing over. You will feel the restraint of Knowledge here if you are sensitive to its presence.

You can become addicted to certain kinds of things—things you do not need, things that are unnatural for you, things that weaken you. These are substances, these are drugs, these are possessions, these are forms of technology, these are forms of stimulation—there are so many inducements in the Greater Community to entice beings who are far stronger than you mentally into wanting these things or to becoming dependent upon them. Even the illegal trade in your part of the universe caters to this to a very great degree.

The hazards are greater here, you see, but it is still the power and presence of Knowledge that protects you, that enables you to recognize a form of persuasion and not to give in to it. Knowledge is not giving in to it, and if Knowledge is your ground of being, then you will not give into it. But if Knowledge is a distant voice in your mind, then other forces are controlling you. Then you are vulnerable.

The Greater Community is an immense environment of persuasion regarding interaction between races. Nations that have chosen to live without freedom and to control their people and to restrict their movements, their thinking and their awareness must exercise great degrees of control. They must prevent their citizens from gaining access to Knowledge. For Knowledge is the beginning of one's liberation. It does not pay homage to the powers of the state or to religious institutions or to the persuasions of commercial interests or to the threat of exclusion, even the threat of death.

Knowledge is the most powerful force in the universe, and those who seek to avoid it must avoid it at all costs, particularly if they are the leaders of nations or have tremendous wealth or power and influence in the realms of trade and commerce. Knowledge is revolutionary because it is inherent within the individual. To prevent people from being aware of it, you must consume their attention with other things, with fear and with desire, with conformity, with work—with overwhelming work.

It is a great challenge, then, to keep Knowledge alive within a nation, particularly a nation that is moving away from Knowledge in its emphasis. It is the most precious thing to keep Knowledge alive within your world as you prepare to face the Great Waves of change that are coming to your world and as you prepare, unknowingly, for your future within a Greater Community.

People think Knowledge is weak. It is just a feeling; it is just a thought. But it represents the core of your strength, the power of your integrity and the source of your true discernment. Lose this and no matter how strong you think you are, no matter how controlling you try to be, you will be easily persuaded and have already been easily persuaded. This is why Knowledge must be a great emphasis, and this is why it represents the core spirituality in the universe.

Regarding the reality of contact with other nations, the power of Knowledge is the power to see beyond deception, the power to block inquiry, the power to know things that others cannot see, the power to maintain your integrity, the power to discern danger and the power to recognize a true friend and ally. It is this greater discernment that must be brought to bear in the complexity of communications, negotiations and trade within the Greater Community, where Knowledge is rare and unknown, but to a relatively small number of individuals.

Because Knowledge has been placed within each person, it represents a potential for freedom and strength. Knowledge also has power in the mental environment—the power to inspire others, the power of inspiration. It is amazingly effective if an individual even has a little opening in their mind. This is what moves you when you hear someone else's genuine experience. This is what moves you when others share an experience of being moved themselves. This is what gives you the power of empathy. This is what enables you to experience another's experience. This is what enables you to value others who seem to be different from you and to value things beyond the persuasions of wealth, power and attraction—things ineffable, but permanent and potent.

In the mental environment, you have to learn how to shield your mind. Others who have skill here are able to do this—to block intrusions, to block potent forces, to block thought forms that are projected at one and to maintain one's awareness if one is being affected by technology, by forms of radiation, by scanning and things of this nature.

In an environment of greater influence and greater deception, the need for Knowledge is even greater and more significant. As it is today, an individual from a Greater Community could simply dominate your mind. You would think what they want you to think. They would put thoughts in your mind and you would say them. They would stimulate your feelings, your desires or your fears, not by providing images, but simply through power in the mental environment.

If you are able to have an objective experience of your mind, to realize that who you are is not your mind and to objectify your experience within your mind, you can resist these temptations. But you must have another foundation within yourself, a foundation of Knowledge. Then you can build a shield; then you can prevent penetration; then you can disassociate yourself from the part of your mind that is being stimulated. Even if someone were to paralyze you physically so that you could not move, you could still see and know them and discern them. Even if there were an attempt to erase your memory, as often occurs to people taken against their will by forces who are intervening in the world, you could recall your memory through Knowledge.

It is only the surface of your mind that can be controlled here.

But if you live at the surface and if you identify yourself with your ideas at the surface of your mind, then you will feel that you yourself are being controlled and manipulated. And you will be controlled and manipulated by the influences of friends and family, by commercial forces and by the bombardment of the images you experience as soon as you leave your home.

It is a very important thing, then, to understand the skill that has been developed in the mental environment and to recognize that it is the power of Knowledge that can offset it. You can create counter movements in the mental environment, but it takes tremendous energy, and it is not expected that you have developed such skills. But you need to learn how to shield your mind, particularly if you are near a Greater Community presence. Since the Greater Community has a presence in your world, it is something you may well encounter.

This is what enables you to be without fear when facing danger. This is what enables you to respond in an emergency, to do what is needed and appropriate without concern for yourself. This is the power of Knowledge. This is what enables you to mount tremendous effort under dire circumstances to save someone's life, to save your life or to do something else that is required for the well-being of others. This is the power of Knowledge. It is what takes you beyond your mind, your thinking, your conditioning and the influence of the mental environment that you live in. This is the power of Knowledge. Others are inspired in spite of their ideas and beliefs, in spite of their fears because Knowledge has touched them. Knowledge within you has touched them.

When you read the Briefings that have been sent from the allies of humanity, you can recognize the intention behind their communications. It is inspiring. Even though their message is difficult to receive and may frighten your mind, it is inspiring because of the Knowledge that it represents and the high regard for humanity this represents.

Your allies will not come to save you. They will not mount an effort to come and save you. They will not betray the secrecy in which they live. They will not enter warfare with other nations on your behalf, and thus destroy everything it has taken for them to develop and to secure their anonymity in the universe. But their inspiration is important because it speaks to your experience. It speaks to Knowledge within you. This is the power of inspiration that can break the hold of the mental environment, that can break the hold of persuasion and that can break the hold of addiction.

If you are to be powerful as a people in the Greater Community, you must look upon these things with great sobriety, not with hope and not with fear, but with Knowledge; not wishing, believing and hoping that everything will turn out right and not with the fear that you will be overwhelmed and cannot respond. It is a different kind of position within yourself. It is a position you take in a house on fire when you must save people. You jump into action. You are not thinking about hope or fear. You are responding to a situation, doing whatever you can to help.

Many people have had this experience. It represents a great state of objectivity. It is not aggressive. It is not destructive. But it is very different from your normal experience. It is this awareness that you must cultivate and support in yourself and others and bring to bear regarding your own understanding of the Greater Community. Without fear, without ambition, without wishful thinking, you can see things clearly—if you look, if you are patient, if you take time. There are no quick answers if you are seeking a deeper experience.

Others will try to influence you, and this influence is underway even at this moment from others in the Greater Community. Will you be able to recognize this influence? Will you be able to recognize with objectivity your own ambitions that will be preyed upon and your own fears that will be preyed upon? If you feel helpless and hopeless, that can be preyed upon. If you secretly desire power and advantage, that can be preyed upon. If you know your own mind, you will see where it is most vulnerable to persuasion and how it responds to this persuasion. You will have a choice then in how to respond, instead of being swept away and finding yourself in a bad state all of a sudden, not knowing how you got there.

Despite the restraints and the difficulties of life in the Greater Community, it is a magnificent creation. What can be seen and experienced there is marvelous. It is nature on a grand scale. It is life expressing itself through an unbelievable panorama of forms and expressions. It is something that will move you deeply, for at a certain level, you are connected to this greater reality, this greater panorama of life. The wisdom that can be gained from the Greater Community is immense and immensely practical and relevant to everything you do and everything you will need to do in the future. The spirituality of the Greater Community is The Way of Knowledge in its purest, most essential form.

This is a source of inspiration because this reminds you that you are greater than your mind or your body, that you have a greater purpose in life and that the events of your life will give you an opportunity to experience and to express this purpose, which would never arise under more quiescent and normal circumstances. Your connection to life beyond this world, which is innate within you, is part of this purpose. You have only to be patient and take the steps to

Knowledge to allow the pieces of the puzzle to come together, as they most surely will. Yet if you are reckless and impatient and try to fill in all of the spaces yourself, you will not see the picture that is waiting for you to discover.

The Greater Community is an experience. It contains everything that is marvelous and everything that is difficult, everything that is true and everything that is deceptive, everything that is wondrous and everything that is dreadful. But you must have great sobriety to approach this—to see with clear eyes, to see without preference, to see what you are looking at and to allow Knowledge to give you the indication if something is good for you or not and how you should respond.

Humanity already has a great promise to be a race that is strong with Knowledge. You have been able to maintain your religious traditions. The Way of Knowledge is not foreign or unknown to you. It is part of your religious traditions. It is part of your worldly experience. In general, primitive peoples are closer to Knowledge than technologically advanced races. But Knowledge is there nonetheless.

If you know you are not your mind and can view your mind objectively and step back from your thoughts and feelings, then you have achieved a greater skill—a skill you will need with ever greater regularity as the world around you becomes more difficult, as the Great Waves of change break upon your shores, creating increasing instability and insecurity. In human societies in the world around you, you will need this inner strength. It must be strong enough not to fail you in your greatest moments of fear, doubt or anxiety. And you must withstand the inducements and seductions of the Greater Community that are being placed upon humanity at this time by small groups of intervening forces.

This is the greatest opportunity you have to fulfill your mission here, to discover your greater power, to discover your connection to God and to find your most potent forms of service to humanity and to the world. In adversity, you find your greatest strengths. Your higher purpose is called out of you facing situations you would rather avoid. If you seek only pleasure, comfort and security, what is great within you will not emerge, and you will neglect or avoid those circumstances that could call it forth from you. You will succumb to the mental environment and to the persuasions that exist there, following the expectations and demands of others, not even knowing where your thoughts come from and how they are being influenced.

Your practice in dealing with the Greater Community happens right in your own home, in your sphere of friends, in your sphere of relationships. Where you work, wherever you participate with others, there is a unique mental environment in that situation. Every home has a unique mental environment because it is created by the intelligences that reside there. It is a part of life. It is an integral part of life. And it becomes a greater part of your life as you advance and progress and gain an understanding of the Greater Community all around you.

The Presence of these Unseen Forces
gathers where Knowledge is being
experienced and expressed.

Those who have become close to Knowledge,
who have taken the steps to Knowledge,
they will feel this Presence.

CHAPTER 11

Unseen Forces in the Universe

Serving all of manifest life in the universe, there are Unseen Forces. These Unseen Forces speak for what is good and necessary. They serve as a source of encouragement and reinforcement for those who are beginning to experience and recognize Knowledge, the great spiritual inheritance that the Creator of all life has imbued in all intelligent races.

The Presence of these Unseen Forces gathers where Knowledge is being experienced and expressed. They constitute the Angelic Presence that is assigned to each world, and it includes those who assist this Angelic Presence as well. Even in worlds where there is no religious tradition that are governed by strict dictatorial forms of government, even there, there is an Angelic Presence. Wherever intelligent life has originated or has migrated, you will find this Presence.

Those who have become close to Knowledge, who have taken the steps to Knowledge, they will feel this Presence. This Presence can guide them and instruct them, warn them and prepare them. Yet because it is beyond the visible range and even beyond the realm of the mental environment, these Unseen Forces can exist without the awareness of those races they are meant to serve. Even those individuals who are developed and skilled in insight and clairvoyance, even they will not be able to discern, with any accuracy, the presence and the work of these Unseen Forces.

In many places in the Greater Community, they are called the Unseen Ones. Their Presence indicates that God has a Plan. The Plan for the universe is to reclaim the separated through Knowledge and, through these individuals, to provide unique forms of contribution for the well-being and the advancement of their races. Here progress is made in terms of the realization of Knowledge, ending the separation between those who are in the physical universe and the Source of all life and, at the same time, providing meaningful and essential contribution to science, philosophy, literature, ethics and so forth.

There is the Presence everywhere where intelligent life exists. In the very primitive societies all the way up to the most technologically advanced societies, this Presence exists. It is not just the Presence of God, which is everywhere. These Forces are here to serve specific goals, to reach certain individuals and to assist those who are beginning to experience the light of inner freedom and those who are beginning to experience a greater Presence around them.

Therefore, practicing engagement with this Presence, or you could say "practicing the Presence," is one of the fundamental practices in The Way of Knowledge. Even in religious traditions where Knowledge is recognized and honored, Practicing the Presence is a very universal form of spiritual practice. The Way of Knowledge represents a universal form of religious experience and spiritual emphasis, an emphasis that can be translated through any culture that has the freedom and the awareness to experience it and to express it.

This awareness of Unseen Forces is very important. These Unseen Forces are regarded in many different ways, as the Great Powers, for example, or as an Angelic Presence. Through many different languages and understandings, the spiritual Divine Presence in life becomes universally acknowledged by those who are experiencing and practicing The Way of Knowledge within their own respective worlds.

These Unseen Forces have a goal. The goal is to ignite a deeper intelligence within individuals, the intelligence of Knowledge, and through this intelligence to contribute to the well-being, protection and advancement of their peoples and nations.

Yet the opposition to these Forces is significant, for strict hierarchical societies are afraid of any other source of authority functioning in the individual's life. Here practices such as practicing the Presence, practicing inner silence, inner listening or receiving communications from Unseen Forces are greatly repressed. Seen now as a danger not only to the allegiance of the individual but to the security of the state, such practices are strictly prohibited, and in many places there is a great effort to discern their presence and to eradicate those who participate in them.

The fact that freedom is so rare in the Greater Community indicates and illustrates in innumerable ways the fundamental dilemmas of living in physical existence and the problems of power and authority in manifest life. Those powers and authorities that are not aware of Knowledge or who do not honor the existence of Knowledge will tend to oppose it and regard it as a threat to their authority and to the stability of the nations that they govern. This again is why the pure practice in The Way of Knowledge is usually done in a clandestine manner, in secrecy, through various networks of individuals who are able to communicate with one another beyond the scrutiny of their own governments and political organizations.

Here you have an immense contrast between the difficulties and restraints of mundane life and the constant presence of Unseen Forces serving a greater plan and purpose for God. This also reflects upon an understanding of how the universe works, for God has set in motion the forces of evolution and the forces of natural selection. Within this pattern, there is tremendous variability. It appears to be chaotic, but within this overall pattern, there is a natural progression towards complexity and ultimately towards Knowledge for intelligent races who have self-awareness.

Here you have overlapping two different kinds of patterns. You have the pattern of evolution that humanity is just beginning to comprehend, and then you have a pattern of evolution regarding awareness and spirituality that has been set in motion and is supported by Unseen Spiritual Forces wherever intelligent life exists. One seems random and chaotic; the other is organized and purposeful. It depends upon the level of understanding and the level of reality which you will consider.

Someone who is only looking at the physiological and biological patterns of existence will see the evolutionary track. They will see the aspects of evolution that are beginning to be understood within your own world. This pattern will seem variable and chaotic, without any notion of intelligent design behind it. But overlapping this, there is an evolutionary pattern of intelligent design, an evolution for intelligent races towards self-awareness and the discovery of Knowledge.

This is a source of great confusion amongst the scientists of your world and indeed the scientists of many worlds. You have overlapping realities here. They are functioning in the same environment. One is discernible through your senses. The other is discernible through a deeper sense within you. One is visible and tangible. The other is invisible, but so very present. You therefore need to allow for this seeming contradiction of these two overlapping realities to exist in order to appreciate and to begin to comprehend the nature of

God's work in the universe.

In societies where Knowledge is practiced, either with official support or in a secret manner, the acknowledgment of this greater evolution is universally recognized. This acknowledgment has its own patterns and its own progress that is both tailored to the individual and that has certain basic thresholds as well. Fundamental within it is the realization that one's true reality is not defined by one's mental state or one's spectrum of ideas or beliefs, that one is not one's mind and that one is connected to a greater intelligence and a greater reality.

You can begin to see here again why Knowledge is such a threat, for a person who is engaged with this reality begins to differentiate from the normal structured patterns of life and is now able to function more creatively with greater discernment. Their mind is free from the governance of societies, political structures, ideology or nationalism. They are now a free-thinking individual, an individual who is guided by an Unseen Power that is beyond the grasp and comprehension of the state and beyond the grasp and comprehension of many religious institutions as well. Governed now by a Greater Power, the individual begins to demonstrate abilities that will not be found by their compatriots and by their peers—a power that is uniform, that is not chaotic, a power that is vastly intelligent and compassionate. This represents the motivation behind all true reformers. Every nation has reformers or at least the inspiration towards reform, progress and advancement.

Here, essentially, you have competing powers. You have the power of the state and the power of large commercial forces that are focused on stability, security and continuity, wealth and power. Then you have another source of authority, the power and presence of Knowledge in the individual supported by the Unseen Forces—the Angelic Presence, if you will. Therefore, there is a competition in the universe—a competition between governance, control and conformity and a greater Spiritual Power that seeks to liberate, that seeks to differentiate and that seeks to express itself through the individual to demonstrate the beneficence of the greater Spiritual Power in the universe.

In the Greater Community, there is the recognition of one God, but in some places this is altered to include many gods serving one God. In other places, there is not the recognition of one God, only of the Spiritual Power and Presence, depending upon the theological emphasis and orientation of the peoples involved. The variability here is significant, but the essence is the same. That is why Knowledge in one world is like Knowledge in another world, functioning beyond and within the great variability of cultures, ethics and orientations that exist between different worlds and nations.

The power of unity in Knowledge is very significant. Knowledge does not condone war or conflict, unethical competition or deception or manipulation. It is entirely just. It is entirely uniform, so much so that individuals from opposite ends of the entire galaxy could recognize and communicate to each other through Knowledge, should they ever have the opportunity to meet.

A nation governed by the inspiration of Knowledge would be entirely different in essence and in emphasis, in structure and in social orientation from a society governed strictly by law and dominance. That is why there is a great difference between a nation that is free and a nation that is not free. You can see the manifestations of this within your own world between nations that have greater freedom and nations that have little or no freedom. You can see the corruptions of power. You can see the emphasis on self-perpetuation that exists in large institutions and how this has a limiting and corrupting influence upon its participants.

Everything We are telling you is within the realm of your experience here in the world. For the laws of nature are the laws of nature, and the Plan and Presence of God and of Unseen Forces are uniform throughout the universe. So while races and nations may appear to be entirely different from one another, they are bound to the same physical reality, and they are influenced by an overseeing reality that is divine and spiritual in nature.

That is why the importance of spirituality and spiritual practice that are translatable is so very significant. This speaks of a greater union in the universe—a union beyond separation, a union beyond the difference of environment and appearance, orientation and belief. Here you see what is powerful, what is consistent, what is uniform and what does not change with time and place, situations and circumstances.

Yet even here Knowledge functions differently in different environments, giving guidance specifically to individuals that is unique to their circumstances and their needs. This demonstrates God working in time through individuals, for this is how God works in the universe. This is a very important understanding. While your emphasis may be on what is different and unique that separates individuals and nations from one another, both nature, which has been created by God, and the Spiritual Presence, which has been set in motion by God as a unifying force, represent the real powers in the universe—the power of nature and the power of God's Plan and Presence.

It is as if God's Plan is a balance and an antidote to the chaotic nature of the natural world. The randomness that one finds in nature and its evolution of life forms is balanced by this constant focus on spiritual awareness amongst intelligent races. Here intelligent life has a deeper instinct to return to the one God, to the One Reality beyond all separation, distinction and disassociation.

It is these overlapping realities, then, that produce a seeming contradiction, but which in fact represent a balanced union—a union between that which grows, changes and evolves and that which is constant and continuous, unchanging and yet adapting itself to changing circumstances. To be really successful in life, you must be able to grasp these two realities and be a bridge between them, treating the mundane natural world at the level which it exists and treating the overlaying Presence, the Divine Presence, as it exists—recognizing randomness, change and variability in one and the continuity, consistency and eternal nature of the other. One is changing; the other is eternal. One is temporary; the other is permanent. The entire universe is temporary; God's Presence is permanent.

To have a connection to these two realities—to see how they interact and to allow them to interact through one's self—represents real spiritual advancement throughout the Greater Community. Here the debate between randomness and intelligent design comes to an end, for they are both true. Here the laws of nature are honored as a vehicle for growth and advancement from simplicity to complexity, allowing the universe to move and to change and allowing life forms to adapt and to evolve. Yet there is another reality that exists and permeates this physical reality that is consistent and constant. Though this other reality adapts itself to changing circumstances to serve manifest life, it itself is completely uniform. It represents part of God's Creation that is not entered into form, that has not entered into a process of change, growth and decay.

This represents the focus of spiritual awareness in the universe. While all peoples and nations must function within the natural world—a world where resources must be obtained, where stability must be established, an environment where survival and continuity are ever present and ever important—there is a greater unifying Power and Presence, unseen and unrecognized except by those who have begun to respond to the power of inspiration within themselves. This Power and Presence are always there, always beckoning intelligent life toward a greater union and a greater experience of Knowledge. This appears to be a great contradiction until you begin to recognize how these two realities complement each other and why the Spiritual Presence and the Unseen Forces in the universe represent the real hope for advancement and redemption for intelligent life everywhere.

The great variability of life in the universe represents the evolution of life at all levels and in countless forms and countless variety, where the distinctions are endless, where the environments are different and where the circumstances are different. Here one could imagine endless variability. But then there is this unifying reality, this Unseen Presence, Unseen Forces working on behalf of Knowledge, working according to a Plan drawn and directed by the same central Power, creating a competition for authority within the individual between the dominance of the mind and the vast intelligence of the Spirit.

Races that have become spiritually advanced have begun to recognize their dual nature and this dual reality that seem to be a contradiction but which, in reality, serve life. While life must evolve and change, it must also be redeemed. While there is countless variety and variability through the evolutionary patterns of adaptation and survival, there is a greater union that exists that is consistent and that does not function in opposition to itself. While there is the inclination for separation and distinction within the individual, there is this greater power of union that unites the individual with others and with life everywhere.

If this overlying reality were not present, the evolution of life would be a constant process of disassociation. Life would become ever more distinct and particular. Variety would continue within the constraints of one's environment. There would be no ability to connect or to deeply associate between races in the universe. Beyond the prospects of trade and commerce, there would be no way for races to truly resonate with one another. The possibility for peaceful relations would be greatly diminished as a result. And religion would never emerge beyond superstition and beyond the rituals of the state or social pageantry to have any greater meaning whatsoever.

Therefore, you have Unseen Forces unifying the universe, providing the basis of unity for life beyond the complexity and variability of life, and with it the promise of redemption for the individual to regain an essential relationship with others and with life itself. Without this, there would only be an awareness of physical life. There would only be the emphasis on separation, on distinction, on disassociation, on conflict, on competition, on dominance, on persuasion, on manipulation and all the forces and activities that are so prevalent throughout the universe.

Without these Unseen Forces, the best that nations could achieve would be an economic interdependence and a state of compromise with one another—a compromise that would always be fragile and subject to change, threatened by difficulty, unstable, uncertain and so forth. Without this Unseen Presence, the evolution of life with an emphasis on stability and security would become so anti-progressive, so conservative, so self-reinforcing that awareness and even technology itself would never really advance. Nations seeking a steady state, seeking a permanent stability, would become so resistant to change, to advancement and to innovation that the progress of intelligent life would be stifled.

You have seen in the history of your own world that civilizations that were the most long lasting tended to be extremely conservative and resistant to change and innovation. Consider this on a larger scale and you can see how nations tend towards this pattern as their emphasis on stability and security increases. But there is a Greater Force in the universe that promotes innovation, promotes reform, promotes advancement, promotes creativity and promotes the freedom to make all of these activities possible.

You do not realize how important these Unseen Forces are in your own personal awareness, the advancement and the possibility for true accomplishment and the establishment of peace and cooperation for your own race. You may think that it is all about negotiations and compromise, war and conflict, struggles for power between different groups and different interests, different tribes and different nations—a constant battle, a constant conflict occurring at every different level.

Without this Greater Presence, life is only a struggle—a struggle to achieve and a struggle to defend what one has, a struggle against endless threats and forces of competition, a struggle against death and destruction, a struggle for wealth and power and a struggle to maintain wealth and power. Without this great Spiritual Presence and these Unseen Forces in the universe, life degrades into a miserable state. Even if one has advanced technology and many conveniences, even if one's nation has been able to establish a continuous state of stability, even if there is social freedom, life becomes only a struggle, a dilemma, a stressful existence and an empty experience.

You value change in the physical reality, but you do not yet see the power of continuity that exists in the deeper well of your mind, in the deeper intelligence within you and in the presence of this intelligence in life. It exists throughout the universe. It is not changed by the change of culture or environment or by the evolution of one race or its social structure, its fantasies and mythologies, its political ideologies or even its religious beliefs.

Everything that is innovative comes from the Mystery of your life. All inspiration comes from the Mystery. All new ideas come from the Mystery. The quest for freedom comes from the Mystery. The quest for justice comes from the Mystery. The quest for union and for peace comes from the Mystery. The recognition between two individuals or even between two worlds comes from the Mystery. It is this Mystery that exists alongside the manifestation of life that gives life all of its value, its meaning, its purpose and its true destiny.

Experience the Mystery and you will see the manifestation in a different way. If you see only the manifestation, you will miss all that the Mystery has to offer you—its guidance, its protection and its ability to take your life into a greater level of service and meaning. These are the Unseen Forces in the universe, permeating and existing alongside the natural world and the existence of nature.

To understand what God
is doing in your world,
you must understand what God
is doing in the entire universe.

CHAPTER 12

Creation and Destiny

The physical universe began with the Separation long ago, even before the beginnings of the physical manifestation of the universe itself. The Separation represents part of God's Creation that has disassociated itself, seeking a separate existence. So the Creator of all life set into motion all of the forces, the geologic forces, to establish a great physical environment in which the separated could evolve and live. The Creator of all life established the processes of evolution and random selection so that the forms of life could evolve in interaction with their environments, the multiple environments that would be produced, and so that the universe essentially would run itself mechanically according to these laws.

All the laws of nature were set in motion. All the laws of physics and chemistry were set in motion so the universe—the physical universe, the manifest universe—could be self-maintaining, self-sustaining and run upon its own so that God would not have to hold everything in the balance. Here the universe would expand, leading to unique opportunities for life to emerge and to exist in primitive forms for long periods until such time as more advanced levels of life could be established, where the separated could enter into these life forms and live in habitable worlds surrounded by other forms of intelligent life.

The universe then is an environment in which life evolves and can exist in many places, in many forms and in many expressions. That part of God's Creation of life now seeking to distinguish itself and to take separate forms would have a great environment in which to live and grow; to face the reality of change, growth and decay; and to have the laws of nature to establish the parameters of life and the vast evolution of life so that many Earths that will be habitable could be filled with a variety of creatures, plants and natural environments.

Life in this physical universe would be temporary. One could not sustain one's life in the manifest form forever. And those who are separated who seek to remain separated would be bound to this reality, would have to function within this reality, would be able to be creative in this reality up to a point, would have to seek resources and would have to face the dilemma of living in separation from God.

The problem of survival, the problem of competition, the problem of illness and exposure to other life forms, the problem of facing difficult environments, the problem of getting along with others in a separated state and the evolutionary track of going from being very primitive peoples into greater dimensions of social manifestation and development, facing competition from other groups within one world and ultimately facing competition from groups from other worlds—this is the arena that God has created for that part of Creation, which is only a small part of Creation, to have a place to live and to grow, to make decisions and to redeem themselves through the discovery of Knowledge and by making contributions to their own peoples and worlds.

This is an oversimplification of the reality of life and the intention of life. But, nonetheless, it states the fundamental principles upon which one can begin to comprehend one's own dual nature—one's physical nature as a creature living in a world, living in an environment, and one's Divine nature—and how different they are from one another.

Yet one is ultimately meant to serve the other. Your physical nature is meant to serve your spiritual nature. Your greater purpose for being in manifest life, a purpose that God has given you, is to provide a unique service to your race while you are here in your temporary existence in the physical reality.

This is the theology of life in the universe. It is not the theology of life for one race alone. Humanity's understanding of its Divine nature and purpose, therefore, is limited because it is only seen within the context of one race—within the context of your own history and series of events, within the context of your own nature and temperament, your own strengths and weaknesses, social structures and inclinations.

Truly, to understand what God is doing in your world, you must understand what God is doing in the entire universe, within the Greater Community of life in which you live. That is why this greater understanding is so essential. When you bring this understanding back to your own existence, it begins to add tremendous clarity to your nature, to your circumstances, to your understanding of yourself and other people, to your inclinations and to an awareness that the Deeper Power exists within you—a power to guide you, to protect you and to lead you to a great accomplishment in life. In a sense, you need a Greater Community understanding to have the best opportunity to see clearly your dual nature and how your physical life is meant to serve the spiritual reality.

While you and others have sought refuge in separation to experience yourself as creators, God has given you another purpose for being in manifest life. You seek separation, but God has given you another purpose and has embedded this purpose within you. For while you attempt to be separate from God, you cannot really be separate from God because the core of your reality and identity is Knowledge itself, which God has created, which could never be separate from its Source.

Perhaps this will be confusing at first, but once you see it, it will be clear as day. Once you gain a vantage point on the mountain of life where you see beyond the trees, the hills and the valleys, you will see the panorama clearly. That is why taking the steps to Knowledge—returning to Knowledge, gaining an awareness and experience of Knowledge—is like climbing a great mountain. At the beginning, you cannot see anything. It is very confusing. And the mountain looks small and easily achievable. But that is because you only see its lower flanks and cannot see its immensity and the great journey it will take to reach its higher vantage points.

People want an easy understanding without making the journey, but the journey is everything. The journey is where you learn to become strong, where your objectives become clear, where you learn the proper criteria for relationships, where you learn what you can take with you and who can journey with you up this mountain, where you gain wisdom and perspective and where you clarify your own intentions and resolve your own conflicts. This is where you leave behind your past of living in confusion, dominated by the social forces around you and dominated by the mental environment in which you live.

Not everyone can travel with you up this mountain, and you cannot take everything with you. You cannot take all of your beliefs, attitudes and assumptions with you. For as you go, things are set aside, eventually only leaving you with what is essential for your journey. When you live with what is essential, you will recognize and feel that your life is essential. Likewise, if you surround your life with what is not essential, you will feel your life is not essential.

Understanding Creation at the level of the Greater Community, then, gives immense clarity to your life and experience of purpose here in this moment, within the limited circumstances of your existence. Here you learn to see the difference between what is different and what is the same and the difference between the great variability of manifest life and the singularity, clarity and union of your inner spiritual life. Here you recognize the great hierarchy of your life, where your body serves your mind and your mind serves your Spirit.

As you translate this into your experience of being in the world, it clears up so much confusion and uncertainty, conflict and self-denial. For you are here now to serve a world in need. That is your greater purpose. Yet only Knowledge in you knows what that service is, where it can be rendered most fully, the other individuals you will need to unite with in order to provide this service, who your essential relationships will be with, what the criteria for those relationships are and how you will be able to unite yourself into one person with one deeper reference point within yourself, instead of being a person who is pulled here and there by many different forces of persuasion.

Only God knows how to redeem you. Only the Plan that has been set in motion can sustain you and unite you and give you the experience of having a greater purpose. All individuals in manifest life have been given a greater purpose, though a very small percentage of them have ever discovered this. A large part of that has to do with the environments in which they live. If you live in a highly structured, highly technological civilization, the possibility for you to experience Knowledge as well as express Knowledge is very small.

Your learning environment, then, can be extremely restrictive. And here again you do not realize the great advantage you have living in a world of relative freedom, a world that has a diversity of religious expressions and experiences. You look at your disadvantages and your weaknesses, but you do not realize, in contrast with the Greater Community in which you live, what immense strengths and opportunities you have. You may lose these opportunities, which you are at risk of doing in facing the great changes that are coming to your environment—the great change coming as a result of your diminishing resources, your ability to grow food, the distribution of water and the risk of conflict and war.

These things will show you how precious your advantages are and how much you and others have neglected them previously—failing to see their great importance, failing to understand how rare they are in a universe where intelligent life more often than not tends toward uniformity and control in environments where there is no freedom, environments where the possibility to discover the power and presence of these Unseen Forces in the universe and the reality of Knowledge within oneself is extremely remote.

Time is not an issue for God, but it is an issue for those who live in manifest life. Time here can be equated to suffering. The time it takes you to come to terms with who and what you really are can be measured in terms of suffering—confusion, denial, self-hatred, depression, addiction and immense errors in life. You are in the wrong place with the wrong people doing the wrong things—things that are harmful, things that are setting you back, things that are holding you back. If you had any sense of your true nature and purpose, you would see how utterly destructive this is, how pointless this is, how in the end it will not amount to anything and how easy it would be for you who have these freedoms in this world to waste this great opportunity you have in living in an emerging world such as this—where there is still freedom of personal expression, where there is a diversity of religious expression and where the opportunity to gain access to Knowledge is still very great.

The shock of learning about life in the Greater Community is the shock of learning how little freedom there is, how oppressive so many nations in the universe are and how devoid of inspiration this existence is for their citizens. Still isolated within your own world, you cannot yet appreciate these things from your own experience. But you can look at your world and see the nations where people's freedoms are restricted and recognize how this could become ever more extreme within a Greater Community environment.

Humanity is heading towards resource depletion. What you need to understand is that every advanced nation in the universe is facing resource depletion. That is the threat that hangs over them. That is their condition. They have generated technological innovation to deal with this problem, but technology requires resources as well. To lose your self-sufficiency in this regard, you cast yourself into a very difficult set of circumstances where your life and your circumstances will be controlled by others—others who are providing you resources that you cannot provide yourself. This establishes nations that are uniform and oppressive, where everyone must think the same and do what they are told. This is the norm among advanced nations in the universe, while free nations, nations that value the creativity and the possibility for the individual, are extremely rare in contrast.

This is not what you want to hear, but it is what you must hear, or you will not understand life beyond your borders. You will not understand the progress of life. You will not understand the difficult circumstances that exist in the Greater Community. And you will not understand the nature and intentions of those who are visiting your world now, why their objectives are what they are and what motivates them.

God has a Greater Plan. You cannot understand it. You can only participate within it. If you participate within it, it will become evident that there is a Plan, that there are Greater Forces beyond your visual range who are assisting you and that there is a great movement not only within your world, but throughout the universe—a movement towards resolution, a movement towards ending separation, a movement of redemption to discover the experience of Knowledge. This is universal.

The problems that you face in this world are the same problems that individuals in all worlds face. The need for freedom is a need that is experienced everywhere—the freedom to think, the freedom to act, the freedom to move around, the freedom to discern one's own inner direction, the freedom to pursue that direction, the freedom to create, the freedom to discover one's greater purpose and to express that purpose—without oppression, without condemnation and without subjugation.

Do not, then, look at the Greater Community as a vast resource awaiting your exploration. Do not think that it is a big playground where you can go and enjoy recreation, traveling around as a tourist, having adventures and so forth. These motivations all demonstrate that you do not take your life seriously and do not recognize the greater opportunity your life is giving you to discover your deeper inclinations and to find a greater fulfillment here.

It is the destiny of all intelligent life to rediscover Knowledge and to return to God through service to their worlds. This is what ends the Separation. For if you are acknowledging, experiencing and following the deeper Knowledge, it means that you are experiencing a relationship with the Divine. You are following the Divine Will. God is speaking to you through Knowledge. You are following this in very specific ways, leading you to do very specific things, to engage with certain individuals and to establish relationships that represent a higher purpose in life. You are still an individual. You are still living in your respective world. You are still facing the challenges of your circumstances. You are still facing the limits of freedom that exist within your world. But now you are connected to something deeper and far more profound—something that brings unity and clarity to your life and relief from your own inner conflicts and contradictions.

The universe then becomes a place where you can contribute. Here contribution becomes your emphasis as you experience a greater fulfillment through service and giving to others, giving from this well of Knowledge that is serving you, that is nourishing you and that is providing you a direction to follow in life, a direction that you are meant to follow.

There is a Presence with you. It is the unseen Presence. It is here to encourage you, to reinforce you, not to rule you. You must find Knowledge to guide you. No spiritual force will rule you. It is Knowledge that you must find and follow. This gives you back to yourself. This establishes your authority in life. This gives you power. This gives you certainty. This gives you direction. And it also connects you to your Source. It gives you fundamentally what you need as an individual, and it begins to end separation in your life.

Here you look at the physical universe as a temporary place. No longer is it so overwhelming. No longer is it so threatening. You are able to travel through it guided by a Greater Power, recognizing its opportunities and its limitations, its pleasures and its tragedies, without becoming lost in these things.

You may pray to God for many things, for advantages and for protection, but you have already been given the greatest gift that God will give you—a gift that remains to be discovered, experienced and expressed. That is why God is bringing the most fundamental spiritual practice in life into the world through the form of The Greater Community Way of Knowledge, for you to practice now what others throughout the universe are practicing, to develop the skills that races throughout the universe are developing, to gain a deeper resonance and therefore a deeper connection to life. Perhaps you already feel a connection to life in the universe because you have a Greater Community orientation within yourself. This will bring clarity to this connection, making it meaningful, purposeful and giving it great relevance in your life.

God has given you free will. You may choose what you will follow in your life though few people fully utilize this freedom of choice. In most places in the universe, they do not have or are not granted the freedom of choice. Yet they still have it. Once they realize that Knowledge is a reality, they will have a choice as to whether to recognize it or not, whether to follow it or not, whether to choose to experience it or not and whether to express what it has given them to express or not.

God will not force redemption upon you. That is why there is no Judgment Day in the universe. God knows that without Knowledge to guide you, you would make foolish and often terrible mistakes. God knows that without Knowledge to guide you, your life would be full of error and difficulty, confusion and uncertainty. This is what has created the world that you see. It is the world where individuals are living without the awareness of Knowledge.

That is why even though you have great freedom compared to the Greater Community, within yourself you are not free. You are a slave to your fears. You are a slave to your desires. You are a slave to your social conditioning and to the expectations of others. You are a slave to your need for approval. You are a slave to your fantasies and your desires. You are a slave even to your goals, which often do not reflect your best interests.

What liberates you from this is a Greater Power that has a greater destiny for you. You have important things to do in life, and if you are not doing them, you will feel restless and discontented, confused, uncertain, critical of yourself, critical of other people, constantly complaining and so forth. You have the freedom to choose. The Presence is with you. You can choose to experience it or not. You can choose to come closer to it or not. You can choose to understand it or not.

Many people lose themselves in trivial things or in the particulars of their busy schedules. They stay busy so they will not feel what they know, so they will not come to terms with themselves, running around constantly, preoccupied, consumed, driven, without any deeper sense of where they are or what they are doing.

In the universe, all beings living in separation have a choice whether to continue to live in separation or to begin a pathway of return, a journey of return to their Source. This is the essence of the matter. The distinctive features of a nation's theology, the differences between religions that exist within your world and other worlds, the difference in belief, the difference in pageantry, the difference in worship or spiritual practice—these are not the essence of the matter. These are not very important. What teacher you follow, what avatar or savior you acknowledge, what rendition of salvation you believe in—these are not the essence of the matter.

What returns you to God is to follow what God has put within you to follow, not what you believe. Your beliefs either help you or are hindering you from discovering the power and the presence of Knowledge within yourself.

Groups and nations in the universe have fought endlessly over their religious beliefs. Who is right? Who is righteous? Whose version of reality is the official version of reality? Whose belief in God is more correct than someone else's belief in God? These are all quandaries in the mind, but to the Spirit there is only the One—the power and presence of Knowledge, what God has put in you and in others to guide you and to lead you to a place of contribution that is entirely natural for you, that represents the fulfillment of the deeper need of your soul.

This is the theology of life in the universe. It is applicable to the larger scheme of things beyond your imagination, and it speaks to the most intimate part of your own existence. It is both vast and incalculable and extremely intimate and essential to everything that you do. To recognize the difference between the manifestations of religion and the fundamental practice in The Way of Knowledge is to really separate the essence of the matter from its forms of expression, to see what is essential and to contrast that with what is only circumstantial.

You can practice The Way of Knowledge within a formal religion or outside a formal religion. This is what shows you what is uniform and consistent in the universe and what connects you to other life forms in the universe and to other members of your Spiritual Family, who are not all human beings. This is what will enable you to communicate with other life forms, to recognize them, to understand their intentions, to discern their behavior, to recognize whether they are strong with Knowledge or not, what guides them, what motivates them and what restricts them. This is the essence of your future communications with life in the universe and the essence of your communication with one another here on Earth.

The theology of the universe is really the clarity and the unity of purpose that you serve with all beings guided by Knowledge. As long as you can discern the difference between Knowledge and the manifestations and creations of the mind, you will be able to see this clearly. And it will make all the difference in how you view yourself and your world and your ability to prepare for your future in the Greater Community.

Eventually,
all the religions of the world
will have to reconsider
their beliefs and positions
regarding God
in light of the Greater Community.

CHAPTER 13

GOD OF THE GREATER COMMUNITY

You must ask yourself, "What about God? Where does God stand in this larger panorama of life? And how is the reality of God different from how God is conceived of here on Earth?"

God of the Greater Community is the God of all who live in the Greater Community, not just the God of humanity—this one tiny race in just one small beautiful world, a world that is like a grain of sand on a beach that extends as far as the eye can see. God is the God of other beings, countless beings, countless races of beings very different from you in appearance, in awareness, in attitude and even in the nature of their intelligence. It is their God, too.

What God has created in them and in you is the power and the presence of Knowledge. That is what makes you united with all intelligent life in the universe. But God has also created the evolution that has created your unique physical vehicles—your body and your intellect. God has set into motion the forces that have established these as well.

This is a very different perspective than thinking that God is the God of one people, one race or one world. That is like a God of one tribe. And this must be a very small god. If you can imagine God of the entire Greater Community, the author of life everywhere, then you have to reconsider your ideas about God.

Eventually, all the religions of the world will have to reconsider their beliefs and positions regarding God, in light of the Greater Community. This is one of the reasons that religions do not deal with this subject or have been very reluctant regarding it, for it will prove that their assumptions are incomplete or in many cases incorrect. This will require a complete revision of their beliefs and ideology. To think of God as preoccupied with human beings, that God has a plan of salvation just for human beings without including the rest of the universe, must be extremely incomplete. Here your ideas and notions of salvation may prove to be utterly incorrect.

If God is the author of all these other races, then where does humanity stand? People imagine a God based upon their own sense of identity—a God in their image, either a God that has the physical presence of a human being or at least a God that thinks along human lines, a God that has human values, human reason, human aspirations, human judgments—a God like you only much, much more powerful. Yet if God is the God of all these others, then how can one say that God thinks the way people think, that God has the values that people value or that God operates according to principles that people consider to be essential?

Here your whole notion of God becomes challenged and must be deeply reconsidered. In earlier times and even in the minds of many people today, there is only God, humanity and the world, as if the rest of the universe were just scenery for this great drama happening here on Earth and that everything is about humanity and God and this world, about Heaven and Hell for humanity and this world. But in the Greater Community, none of this can be sustained.

Where does humanity stand in relation to other nations, other civilizations, in the estimation of God? Is humanity still important in the universe? Or is humanity just one of countless races that have emerged through the process of evolution in a vast and seemingly impersonal universe?

God of the Greater Community cannot be preoccupied with one race alone. God of the universe cannot manage the affairs or the events of one race alone. To presume that God is governing events in the world is to think of humanity living in isolation. God is too intelligent to have to manage all of your affairs and events. That is why the process of evolution has been set in motion. That is why the laws of nature were set in motion. That is why the entire process of life in the physical reality, as you are now learning to discern it, has been set into motion, so that God does not have to manage everything. The events of your life are either the outcome of the intentions of you or someone else or of simply random motions.

God is like the great attraction calling everyone to return. God has placed Knowledge within all sentient life, which provides a unique plan of fulfillment and redemption for each specific race according to their nature and circumstances. There is a Plan of salvation for every race, and each is slightly different even though the great attraction to God is the same. The Plan of salvation must be beyond human estimation if it is to have such a universal application. Said in the simplest manner, it is that the separated are reclaimed through Knowledge, and Knowledge leads them to fulfill their lives through contribution to others and to the resolution of conflict.

You see, this is a very simple definition, but what does it really mean? How is it really portrayed? How can this be experienced in life? It is completely translatable, of course, from one person to another, from one nation to another and from one world to another. But it requires a very deep well of experience to really understand what this means. You cannot stand at the surface and know what is occurring at the depths.

Your God is the God of the Greater Community, the God of not only the physical universe, but universes in other dimensions. Even here, all of this physical manifestation is but one part of Creation. How could anyone's theories or philosophy account for this? Surely, any attempt to illustrate what Heaven would be like would appear to be extremely childish, even comical, given the scope of God's Creation. And what would a heavenly state be like for sentient beings from other worlds? If you all returned to the same reality, what kind of reality would that be? It would not be a human reality, unless you think the universe is populated by human beings or beings that are human-like. And that is surely not the case. So what would a heavenly state be like if it is not based upon human values, human images and human aspirations? This is why the reality of the Greater Community will require all of Earth's religions to re-evaluate their positions and to greatly modify their ideologies.

You are moving into a much larger context. In a larger context, the meaning of things changes and expands, and many of your former ideas prove to be greatly inadequate. Yet to say that God is unknowable is only true in terms of your ability to create concepts to conceive of God or to give God an image. But God's Will is knowable in your life, and that is the meaning of Knowledge within you. God's Creation in its entirety is not knowable, but your connection to it is.

Here again is the difference between the meaning of the essence and the manifestations of the form. If God can be worshiped in trillions of different ways and they are all appropriate, how can you say on Earth that your path to God is the only way? Since God has initiated all the world's great religions, how can any of them proclaim preeminence over the others, proclaiming that they are the one true path to God? Even within your own world, this becomes untenable, extreme and arrogant.

The God of the Greater Community requires a deep resonance, and this resonance happens at the level of Knowledge. Resonance is not the same as understanding, intellectual understanding. It is more of a deep experience of affinity, of relationship, of being related. No one's intellect in the world and no one's intellect even in the Greater Community can encompass the meaning of this. But the experience is available, and it is there.

There are some people in scientific circles who can only accept God as being nature itself. But nature is a manifestation of form, not essence. And in the Greater Community, it has been discovered that there are two overlapping realities in life: the manifestations of nature—the evolution of worlds, the evolution of life— which seem random and chaotic without specific design or focus, overlaid by the harmonizing power of the Unseen Forces of the universe. This stimulates and directs a deeper Knowledge that resides within all sentient life. This is uniform. This has intelligent design. This is focused and has purpose and direction.

In these two overlapping realities, to say that only one is God is incomplete. If you only say that the world of Spirit is God, then what about the physical universe? Some religions refer to it as only an illusion, something that is impermanent, that is passing, and that once you gain a higher level of consciousness, you do not need to exist in this physical reality. It is like a great phantom that is sustained by all who believe in it and participate in it. But even this is not complete because God has created and set into motion the geologic forces and the evolutionary forces of nature. It is part of God's Creation. It may be changeable and ultimately temporary, but it is still part of God's Creation. You cannot ignore it or deny it without putting yourself in great jeopardy or peril and without losing your sense of responsibility to serve this reality.

To call God nature is only one side of the coin. It is only one part of the reality. This is difficult because nature does seem to be random. It does not seem to have any specific purpose. Even though life forms tend to evolve into greater forms of complexity, nature does not seem to have any discernible purpose. It just exists. God without a purpose is something that people cannot relate to because people need purpose. They need a purpose they can share with God in order to have a relationship with God. You can just exist and God can just exist, but relationships demonstrate purpose, particularly when you are living in a world of change where the meaning and purpose of what you do in life really does matter.

Therefore, the theology of the Greater Community brings together a complete understanding. It is not a rejection of the life of Spirit, and it is not a rejection of the life of the world—the physical or the mundane. It does not reject the evolution of worlds or the evolution of life. And it does not reject a higher power and a greater creation that are beyond the physical realm. It sees unity in this duality.

The picture is complete, for you yourself have a dual nature. You have a nature as a physical creature living in the world, living within the laws of nature and facing the consequences of violating the laws of nature, subject to your environment and your environmental conditions and the need for food, water, shelter, security and companionship.

It also acknowledges that you are a spiritual being, that you function at a higher level and have deeper connections and ultimately a connection to God. You have the ability to see and know things beyond the limits of your senses, you can conceive of the future and the past and you can see the processes of evolution. You have the ability to understand life at a greater level and in a larger dimension. Yet you suffer over your internal conflicts, you are afraid of death and change, you live in anxiety, and your behavior can become erratic and destructive because of the conflicts that exist within you. While you have power, you are also very vulnerable—vulnerable to error, vulnerable to conflict—and are capable of producing great harm.

If you say that God lives in the heavenly realm and then there is the world, they are not the same. If God is the author of both, how can this be? It is full of contradiction. If God is author of the world, then God is the source of everything that happens here—error, destruction, cruelty and war.

In the Greater Community, these tremendous problems in understanding have been surpassed by those who are advanced in The Way of Knowledge, in the realization and experience of spirituality and in recognizing the inherent harmony and purpose of this dual reality that exists within you and throughout the universe. This represents God's Plan, a Plan that is greater than your society or your world, greater than your intellectual capacity. It is arrogant and foolish to assume that you can know God's Plan, even for your own race.

Here there is no arrogance and there is no superstition. Here there is no trying to triumph over others by proclaiming that your path to God is the ultimate and is the only true path. No one with a Greater Community awareness could say such a thing. To return to God through the reclamation of Knowledge can be practiced in trillions of ways. How can you say that your way is the way? Even on

Earth, there is no one way, for God has created different traditions so that everyone would have a way, so that everyone could utilize a pathway that was shared by other people. Yet religion in the hands of people becomes something else. It becomes an institution. It becomes a set of rules. It becomes stories from the past and the worship of individuals. It becomes something else.

Do not worry. The same mistakes are made all over the universe, often with even more terrible consequences. But there is a fundamental harmony and purpose to the physical reality and to the spiritual reality and to their inherent relationship with one another. It ultimately has to do with who you are, where you came from, why you are here and where you are destined to go. This holds true for sentient life everywhere, regardless of the vast differences in appearance, environment and understanding.

Words cannot express this fully. They can only give you an idea. Knowledge within you will carry your life into a new expression. If you can follow this, if you can learn how to follow this and learn the wisdom you will need to carry Knowledge into the world where it is unknown and often unwelcome, this will begin to show you the Power and the Presence working in your life, whether you are religious or not, whether you visit a church or a temple or not. God's Plan is not dependent upon human creations, but it can work through human creations to great benefit. This is bringing a Greater Community understanding into the world. This is correcting many errors and opening many doors.

God of the Greater Community is not in the business of managing everyone's personal life or arranging everyone's personal experience or creating miracles for everyone. God is the great force that attracts. It is those who serve God, both within manifest life and beyond manifest life, who bring this attraction into the awareness of those who live in physical reality and who support such individuals in receiving this experience and allowing it to take hold in their lives.

It is a Plan so vast, so inclusive and so complex that to even try to understand it is a futile exercise. It is more important that you learn The Way of Knowledge and begin to follow Knowledge than try to create a concept of how God is working in the physical universe. Trying to understand that which is beyond understanding is an attractive proposition for many people, but it does not produce valuable results, only confabulated ideas and immense theories. Reality happens at another level, at a deeper level beyond the intellect. You are not here to try to understand God, but to follow what God has placed within you to follow and to receive what God has placed within you to receive.

This is a Greater Community understanding brought into the world. It is direct. It is fundamental. It is without massive complexity. It does not require tremendous sophistication. But it does require wisdom, honesty and a determined approach.

God of the Greater Community does not give special attention to one race. But the Plan does accelerate in races that can begin to gain the freedom to experience and to express Knowledge. This focus accelerates in individuals who can begin to experience and to express Knowledge.

In the Greater Community, there is no Judgment Day. That is a projection of human grievance. People want God to judge others who they themselves cannot stand, who they reject or who they hate. They want God to exact punishment that they are either unwilling or unable to exact themselves. They want justice, and they think they know what justice is and that God will exact this justice for them.

If you do not discover Knowledge and follow Knowledge, you

are stuck in the physical reality—a reality that is essentially difficult to function within, a reality that is extremely problematic, where suffering is inevitable. God does not dispel your grievances, your hostilities and your ignorance. You have to respond to what God has placed within you. You have to follow it. You have to honor it. You have to stay with it and express what it gives you to express and learn how to do this in a complex, conflicted worldly environment.

There is no creation story. There is no final culmination to all experience that you can conceive of. So what, then, is religion without a creation story, without a final culminating experience and without a judgment day? Clearly, if religion is to be of service to its peoples, it must have a deeper foundation. It cannot be a story or a threat.

In the Greater Community, much of this is set aside amongst advanced nations and advanced individuals. No matter what traditions they have come out of, no matter what patterns of thinking they use, ultimately they come to the realization that there is a deeper intelligence within them that they must respond to, that they must discern from other powers around them and even within them, and that they must follow this and express what it gives them to express. This is the essence of spirituality. This is the essence of spiritual practice. This is the absolute core. This is the pearl.

Humanity has not yet developed the self-confidence to practice this on any significant scale, but the opportunity is still there. It is founded on the individual. The wisdom necessary to approach this Knowledge, to experience it, to discern it and to express it is something that humanity has gained. But now there is wisdom from the Greater Community being brought into the world as part of a New Message for humanity.

This wisdom is very important, or you will misconstrue what

is being presented here. You will make serious errors, and you will not find what you are looking for. Thinking that you are following a deeper Knowledge, you will make critical mistakes. This wisdom is important. Much wisdom has been created, established and learned in the Greater Community, in civilizations far older than your own. To bring wisdom from the Greater Community into the world is of immense value then.

God of the Greater Community responds to Knowledge within you and responds to your intention for Knowledge and your experience of Knowledge. If you are destructive or foolish, you will miss your great opportunity. But you are still stuck in the physical reality. Where else can you go? Perhaps your life will become more miserable and more hellish, but you cannot find reprieve from physical life. After a while, you would want reprieve.

That is why physical immortality would be like Hell itself. For Hell is not a terrible place. Hell is a beautiful place where you can never be happy. If Hell were a terrible place, you would want to escape it immediately. But if Hell were a beautiful place, then you would be ambivalent about leaving it. It attracts you. It seduces you. It gives you hope and encouragement and yet limits you. It hurts you, it deprives you and it disappoints you. Hell is a beautiful place where you can never be happy.

This is what keeps people from returning to their Source—trying to find happiness where happiness cannot be found, trying to find meaning where meaning cannot be found, trying to make separation work, trying to make separation fulfilling, trying to fulfill their ambition for separation and their desire for separation.

People gain wealth and power, and they are still unhappy. People achieve their goals, and they are still unhappy. People have all manner of pleasure and advantages, and they are still unhappy. You know of this. It is evident all around you, and yet everyone is still clamoring for the same wealth, beauty and advantages, knowing that it does not really make a difference. It only traps you more in a set of circumstances where you cannot find the source of your fulfillment or inner guidance.

God in the Greater Community is a great attraction. God is not the paternal or tyrannical role that is imagined so much in your world and in so many other worlds as well. There is something subtle and deeper within the person, something beyond the grasp of the intellect, yet something that can be immediately available—a power, a presence, a movement, a motivation. In the Greater Community, it is called Knowledge because it leads to the ability to see, to know and to act with power and integrity.

Religion, when it is distilled down to its essential elements, is about Knowledge and the way to Knowledge. This is what the Greater Community has discovered. This is what humanity must now cultivate and come to understand.

This Power and Presence lives within you. It is within others in the Greater Community. It is almost beyond the reach of the vast majority of beings living in highly technological or repressive societies. But it is still there, nonetheless.

The movement for liberation in the universe is the sharing of Knowledge, the awareness of Knowledge, the experience of Knowledge and the wisdom that must attend the reclamation and expression of Knowledge. This is the great liberation movement in the universe that is going on everywhere, supported by advanced individuals in free nations, assisted by the Unseen Powers in the universe and ultimately directed and focused by the Will, the Power and the attraction of God.

You can experience this without ever having a belief in God, without ever even thinking about God. Eventually, you will think about God because Knowledge will stimulate this awareness. Yet you do not have to have a belief in God or belong to a religion to begin to focus on the reality of the power and purpose of Knowledge. There are many societies in the universe where there are no religions, so religion cannot be a prerequisite. If religion is really focused upon Knowledge, it is extremely important in creating a pathway. But there are many worlds where religion does not exist beyond the worship of the state and the leadership of the state.

The meaning of this for humanity is immense. What will end war? It is adaptation to your circumstances, and it is The Way of Knowledge. What will prepare you for the Greater Community? It is your adaptation to your circumstances, which are now changing, and it is The Way of Knowledge. What will give you strength, clarity and the assurance of freedom in the universe? It is human unity, self-sufficiency, discretion and Knowledge. That is why your focus on these things must not simply be out of a voyeuristic curiosity, but out of a deeper need within yourself. For you are connected to the Greater Community, and that is why you are reading and studying these teachings. This is not an idle curiosity. It represents a deep connection and a fundamental need.

Look at the stars in the sky,
and consider them in light of the Revelation
that has been provided for you here,
and ask yourself,
"Why am I so drawn to this?
Why is this of interest to me?
What is my connection to life
beyond this world?"

CHAPTER 14

Your Relationship with the Greater Community

You have a greater relationship with life beyond your world. You have this greater relationship because it represents both your future and your past. If you are drawn to life in the universe or have an inexplicable interest in the prospects for Contact, or if you are fascinated with the night sky and the recognition that you live in space, it is because you have had past experience in the Greater Community. Perhaps you have lived in many places, and you bring this with you. It is part of your inner nature.

You have relationships abroad that were established perhaps long ago. But they are still real, for genuine relationships do not die when people pass away. They are permanent creations. Therefore, you may have deep relationships with individuals in the Greater Community. You most certainly have deep relationships amongst the Unseen Forces within this world, for they have brought you into this world and helped prepare you for this life experience.

Perhaps you are drawn to a greater Knowledge and Wisdom, Knowledge and Wisdom from the Greater Community, far surpassing anything that has yet been realized or established here on Earth. Perhaps you realize this is part of your greater purpose in some inexplicable way. There is a connection, a relationship, a deep bond that runs beneath the surface of your mind that you experience sharply from time to time. Perhaps reading and studying these teachings arouse a sense of ancient memory in you or a sense of future responsibility.

Whatever is the case, you likely have a relationship and a connection to the Greater Community. This is entirely appropriate, for many people have come into the world at this time who have a very strong connection with the Greater Community. For this is the time that humanity will begin to emerge into the Greater Community and will have to contend with visitation and intervention from the Greater Community—a great turning point that few people recognize.

For those who especially feel that they have a deeper relationship with the Greater Community, it is a time of great significance. It has affected their life journey so far, as if they were living two lives: one a normal life of a person in the world, and the other a greater life, a life that is connected to greater things that are beyond the scope and interest of most people.

You cannot erase the accomplishments and the relationships you have made in the past, in your past life experiences. You bring this forward with you. If much of your history has been beyond this world, then that is part of your gathered wisdom. It is part of your orientation, and it represents part of your network of relationships.

Even the scientists who study the skies and the heavens, even they have a connection to the Greater Community. Perhaps they only consider it within the context of their scientific study, training and interest. But there is a connection there nonetheless—an abiding fascination, a deep interest. They are drawn to it. It is significant for them. Even the person who has no interest in spirituality, even a person whose vision is entirely based on human reason and logic, even they too will be drawn to the Greater Community. Even a person who has had very little history in the Greater Community can be attracted, for it represents your future and your destiny.

The great forces that will shape humanity's future and destiny, the Great Waves of change that are coming to the world—the depletion of your resources, environmental degradation, the heating of the world, the changing circumstances and the growing threat of conflict and war—are but one set of great forces impacting the world. These are forces created largely as a result of the misuse and the overuse of the world.

Yet there are other forces that humanity did not create for itself. For a long time, the Earth has been watched and studied—the advent of humanity has been watched and studied, humanity's technological process and progress have been watched and studied, awaiting the time when the Intervention would be attempted. For this world is a great prize to those few races who are aware of it. Only your potential allies in space do not seek advantage here. Their interest is largely geared toward their own security and their natural inclination to serve a talented and gifted race such as humanity.

You are emerging into a complicated set of affairs, and you must become prepared for this. These teachings and this book represent part of that preparation. The great change that is coming to the world and the world's emergence into the Greater Community represent the two greatest events of your time and could even be considered the greatest events of all time. They are both evolutionary. They both represent a great challenge and threat to humanity's future, freedom and well-being. They both will require a greater maturity and a greater unity amongst the human family.

Here wisdom about the Greater Community will give you wisdom about the world. Learning about the Greater Community will teach you how you must behave in your own world—what you must do, what you must change and what you must undo in order to proceed safely, carefully and with greater discernment.

Everyone has a connection to the direction and the future of the world, so the connection to the Greater Community is enormous. It is vast. Even for those who do not have a strong history there, even they will have to recognize the importance and the significance of this. To realize that you are not alone in the universe or even within your own world is a life-changing recognition. To recognize that visitation is occurring without humanity's awareness and approval is a life-changing realization. Realizing your vulnerability to space and your lack of unity, wisdom and discernment is a life-changing realization.

After this occurs, following these realizations, you can no longer believe that you are living in isolation in the universe. You can no longer tell yourself that you are entirely safe and secure in your world. You can no longer lose yourself fully in your hobbies, in your distractions, in your romances and in your conflicts, for in the back of your mind, there will always be this awareness of the Great Waves of change and the reality that you are living in a universe full of intelligent life.

Even your science is beginning to realize the prevalence of terrestrial worlds and the greater and greater possibility for the evolution of intelligent life, through observation and through increased probability. Your governments know this, and they hide their secrets because what they know would be so immensely shocking to their constituencies, to the public, to people everywhere who are not looking, who are not paying attention and who are still lost in the notion that they are living alone and isolated in a beautiful world, content to think that others cannot reach your shores or that you remain hidden in an empty universe.

These shocking revelations may be disturbing at first. But they also serve as a kind of confirmation of things you have deeply felt and known for a very long time, even reaching back to your childhood. It is a confirmation of a greater wisdom that you possess unknowingly. It reflects the presence of Knowledge within yourself—the great spiritual intelligence that the Creator has bestowed upon you as the greatest possible endowment.

No matter how you regard the Greater Community—whether in the sphere of science, religion, philosophy, sociology or just as a native person of the world—it is the connection here that must be acknowledged. It is no accident you are reading these pages and are the recipient of this rare and unique wisdom. There is no university in the world that could teach you what is presented in these pages. For how could they know?

This must come through a revelation, which in itself is a shocking realization. To be given a window into the larger universe, to have a notion of how other races behave and respond, what their priorities are, the variety of their structures, the rarity of freedom and the power of influence—these are all extremely important for your preparation for the Greater Community. Without knowing what you are preparing for or having any sense of that, how would you ever know what to do? How could you ever establish a pathway of preparation, determine your priorities or see the risks that humanity is taking even at this moment in neglecting this, the greatest of all frontiers?

You would see that to prepare for the Greater Community, war would have to be stopped, people would have to be educated and nations would have to come together to establish plans for mutual benefit and security. The eyes of the world would have to turn outward to watch, to report, to be discerning, to take great care as to who is here and what they are doing. This could no longer be the focus of some secret government group, elite group of scientists or some religious affiliation that has taken an interest in these matters. Everyone should be watching, listening and learning.

To understand religion and spirituality, you must gain a greater panorama of life as your context. Otherwise, your notions of God reflect nothing but a local deity, a kind of god of your region, a primitive god, the way you look at primitive peoples and how they assign the role of deities to all the powerful forces they experience right around themselves. To understand the unity of religions, you would have to at least contemplate the immense diversity of religious expression and practice in the universe. And you would have to consider that if you feel God is real, then your God is the God of this universe and everything in it and its countless forms of intelligent life, all so very different from you.

Everything becomes broadened and expanded here. This is the perfect antidote to eccentricity, to religious fundamentalism, to extreme religious views and to self-righteousness and domination over others. Even in the realm of security, you must begin to think about the security of the whole world and not just the security of your group or nation. From a Greater Community perspective, you are all one people. The differences between you are minor and insignificant compared to the differences between you and everyone else in the universe.

Whether you are the leader of the most powerful nation or the person who is destitute on the street, you all share the same fate and destiny in the Greater Community. You are entering an environment now that will include competition from beyond and influence from beyond. You will have to adapt to this and prepare for this, to learn how to discern this, to gain greater strength and to gain greater cooperation and greater courage. These are exactly the kinds of circumstances that will mature humanity and enable you to outgrow your ancient animosities, your fabulous fantasies and your tragic involvements.

What will advance humanity but having to adapt to a new set of circumstances? This will be true in your world as you have to face the Great Waves of change, adapt to these and alter the nature and the development of human civilization in order to secure your position in the world. And you will have to adapt increasingly to the power, the presence and the influence of the Greater Community. For this you will need clarity, courage, common sense and a New Revelation.

The Creator of all life is creating and sending a New Revelation. And you have allies in the universe who are not allowed to interfere with your world, but who have sent information in the form of a set of Briefings to help prepare humanity to learn how to establish its own rules of engagement with intelligent life and to emphasize the need to establish your own ethics of contact. For you do not want to be the unwitting recipients of whatever anyone else wants to do here. You have a responsibility as the native peoples of the world to establish a boundary to space and to determine who may enter here and under what circumstances, reflecting the will and the awareness of the people of Earth.

Such rules of engagement and boundaries do not now exist, and this is leaving you open and vulnerable. If you could understand the reality of life in the universe and gain vision and wisdom here, you would see how necessary it is for you to establish these boundaries, to give this your great focus, and how much this could generate, out of necessity, a functional human unity and cooperation, despite the grave differences and conflicts that still exist between peoples and nations in the world.

It is those who feel the strongest connection to the Greater Community who will help educate others—providing wisdom, sharing this New Revelation, contemplating what is being presented here, considering its implications for humanity and realizing how weak and divided humanity is in the face of great united forces in the universe. Surely this must prompt an emphasis on human unity and cooperation, the cessation of conflict and the necessity of giving the world's focus, energy and attention to this great matter.

For you will have only one opportunity to emerge into a Greater Community of intelligent life. How this is carried out and the wisdom you can bring to bear here will have great consequences for your future. That is why the New Message and this revelation are being given to warn and to prepare humanity for the two great events of your era—the Great Waves of change and your emergence into a Greater Community of intelligent life.

The world is ripe for visitation. You are vulnerable to persuasion and manipulation. You have not established your rules of engagement nor have you built your boundaries to space. All native peoples must do this in their respective worlds. It is part of their essential responsibility in order to provide stability and security. And you must learn to preserve your resources so that your self-sufficiency can be maintained in the world. For this will be of essential importance in determining the kinds of choices you will have in the future and the prospects for humanity's freedom and sovereignty in this world.

This is challenging and difficult. Many will turn away frightened or go into denial. But you are still connected to the Greater Community. You have allies in the Greater Community. You yourself likely have important relationships out there somewhere, living in the vastness of space. You have a destiny in the Greater Community. If you are aware of this connection, you will see that the Greater Community cannot be avoided or ignored.

It is the greatest event in human history to begin to engage with intelligent life in the universe. But it is not happening on your terms, and you must look at this with clear and objective eyes. You must have this courage and this sobriety.

The Greater Community is a challenging environment. It is not for the faint of heart, for thrill seekers or for those who want to enrich themselves either financially or spiritually. It is a difficult environment, but a magnificent one, and one that provides all of the incentives and requirements for a young emerging race such as humanity to gather itself together, to unite itself, to dedicate itself, to establish itself, to defend itself and to recognize its great value, its great talents and its wisdom that have been accumulated over a very long period of time.

Religion now will have to become a religion that can function within a Greater Community context. It cannot be anchored in the distant past. For the future will overtake the past. The future will be so demanding it will overtake this reference to the past. Your great religious traditions must now accept they are functioning in a Greater Community of intelligent life. And the reality and the understanding of the Divine presence, purpose and activity here on Earth and beyond must be reconsidered. The notions of Heaven and Hell must be reconsidered, or these religions will not be able to survive into the future. They will become increasingly irrelevant to the pressing and important needs of humanity.

It will not be enough to believe in God if you cannot experience and follow what God has placed within you to experience and to follow—a deeper Knowledge and a greater power. Nor can antiquity really teach you what is happening now. Ancient prophecies are not as revealing as current events in this matter. Do not try to connect the future and the past, for you are emerging into a future that will be unlike the past in so very many ways. Your children and their children will be living in a world that your parents could hardly recognize, facing problems and opportunities that were not of concern to even your most recent ancestors.

Look to the present. Look to the future. Watch what is coming over the horizon in your own world. Look at the stars in the sky, and consider them in light of the revelation that has been provided for you here, and ask yourself, "Why am I so drawn to this? Why is this of interest to me? What is my connection to life beyond this world?"

Do not worry about the rest of humanity. It is better to devote your energy to discovering your own inner orientation than it is to criticizing or evaluating others. If you are amongst the first to respond to the reality and the meaning of the Greater Community, then that is your calling. Do not look over your shoulder wanting and expecting others to have the same response, for if you do, you will be disappointed. This is for you and for others like you who are drawn to this greater reality—drawn because it is of the utmost importance, drawn because it has something to do with who you are and why you are here in the world at this time. This is part of the world you have come to serve.

The Greater Community will have ever-greater bearing and significance for the future of humanity and for everything that is occurring in the world today. The Greater Community will determine the requirements that must be placed upon humanity—the requirement to end war, the requirement to establish a greater unity and cooperation, the requirement to establish a boundary to space and discernment in space, the requirement to establish your own rules of engagement and to concern yourselves with what ethics you will bring to bear—considering who you will respond to and how you will respond and what is appropriate and what is inappropriate regarding your engagement with life beyond this world.

Here many more people will have to wake up to the reality that the Earth is being visited and has been for a long time. This cannot be some kind of exciting fascination. It has to be a sober reality. This is part of the condition of the world that you live in now and will be an ever-greater part of your world as you proceed forward.

You who have a Greater Community connection will have to function without agreement from others, without approval from others, even amongst your friends and family. It represents integrity to do this and strength of will and determination. Study what is being revealed in these pages, which are but a part of the Teaching in The Greater Community Way of Knowledge. Learn as much as you can from this revelation, and bring this understanding to bear upon the world you see and the decisions that you feel you must make within your own life.

Do not denigrate humanity. Do not lose faith in human leadership or human institutions. Do not think that humanity cannot meet its great challenges and problems. For if you do, you will have capitulated—you will have capitulated to the will and influence of those who are intervening in the world, and you will have given up before your strengths were ever really tested. Be careful here not to lose faith in yourself and in the human family. With this faith, you can do amazing things. But without it, nothing will be done.

Humanity's future and destiny are within a Greater Community of intelligent life. To know that future, you must learn about the Greater Community. This requires a revelation from beyond the world, from the Creator of all life—a revelation that can only come from a greater Source. This will give you the context in which to understand how to see life in the world and beyond, how to prepare for the future and the great need for a Greater Community education to be given to people everywhere—to average citizens, to people in positions of power and influence, to the scientific community and to the religious community.

In every aspect of human life, a Greater Community understanding and awareness are necessary. This is to prepare humanity for the future and to offset the damaging influences that are being placed upon humanity at this time from intervening races who are seeking to plant ideas in humanity's mind that will undermine its strength, its power and its wisdom. For the world is already being influenced by certain groups from the Greater Community. This is part of your reality now.

Extraterrestrial life will not be some distant fantasy, but a growing reality and a growing emphasis. This will break through the veil of secrecy and ridicule that has surrounded it in the past century to come back into people's minds, conversations and awareness. It is a matter of great importance and great significance. You will have an opportunity to share what you are learning with others, but at the moment there is no real education in the world about the Greater Community. That is why this New Revelation is here. That is why it has called to you.

Many people want answers. They want to know dates, names

and places. But these are meaningless. You will not be able to visit these places for a long time. It is the perspective and understanding that are necessary. Without this, no matter how much information you have, you will still not see the picture clearly. It will still be utterly confusing to you. You will still think it is what you want it to be. Here wisdom will be replaced by fascination, imagination and the projection of one's hopes and beliefs.

Greater Community education must begin. It is natural for those who are connected to the Greater Community to gravitate towards this and to take an interest here. Do not be worried that humanity remains ignorant, for education can spread quickly, for there are many people in the world today who have a Greater Community connection. Once they hear about the Greater Community and learn about the reality of the Greater Community, they will become ignited. It will speak to them, for they are already connected to it.

The great Unseen Spiritual Forces that exist in your world will support this wisdom and this education, for it is vital for humanity's future. It is vital for humanity's education and understanding of what must be done in the world now and where the strength to do what must be done will come from. You are part of the Greater Community. You have always lived in the Greater Community. And now it is time for humanity to emerge into the Greater Community, which was always its destiny.

MESSAGE FROM MARSHALL VIAN SUMMERS

For me, life in the universe is not an idea, a hypothesis or a mere possibility. It is the timeless reality in which every aspect and event of our life, individually and collectively, have taken place.

Beyond the curtain of shining stars above us is the living cosmos, pulsing with movement, change, conflict and relationship. It is a vast community—a billion billion civilizations undergoing a relentless process of birth, growth, maturity and decay, all in an expanding universe whose outer limits none have explored.

This community is governed by the quiet but pervasive Presence of God, who billions of years ago sparked the creation of the manifest universe, offering form and separation to those who desired it. The forces of physics and evolution were set in motion. Life would evolve. And yet upon the creation of this universe, the Creator set in motion a plan to end separation and ultimately reunite all sentient life. This is the universe in which we have always lived. And this has everything to do with who we are, why we are here and the future of our world.

People ask how I know this? I know this because of the New Message from God. For the last 30 years, I have walked a mysterious path, a journey of a thousand thousand steps. This path has led me, and those with me, to the greatest heights, to vistas where our future and destiny could be glimpsed with stunning clarity. It has also led me back down into the gullies and ravines of mundane life, where it was difficult to see the larger scope of the revelation being given to me.

Navigating this landscape of the remarkable and the mundane

has been my journey to take. Yet over countless miles, this mysterious journey has yielded something of the greatest magnitude for you, for others and for all humanity.

A New Message from God has come into the world. This is a once-in-a-millennium event. It has been delivered through me by the Angelic Presence, who have long prepared me to receive it. And today it is enormously vast—over 9000 pages, given over a 30-year period. This revelation from the Creator has been given to rekindle the spiritual intelligence and fire in people everywhere, from all nations and faith traditions. It is also here to warn us of the manifest dangers of our time and to prepare the human family for its destiny within a larger arena of life called the "Greater Community."

Without such a revelation, could we know our place and purpose in the universe? Could we know the destiny of humanity and what awaits us beyond the borders of our world? Could we gain the wisdom, the understanding and the motivation to preserve our world, to end our ceaseless conflicts with each other and to emerge safely and successfully into a non-human universe, a universe which does not have our interests in mind? This could not be achieved without a revelation from God.

This book is part of that revelation. It is the culmination of a communication from beyond the physical realm, revealing things that even with all the best efforts of science and philosophy could not be known within the span of our lives or the lives of our children, if ever.

This book was delivered in a lightning strike over a period of three days in 2008. After its delivery, I was just as stunned, overwhelmed and inspired as those around me who witnessed it. God had revealed the reality and spirituality of life in the universe.

Yet in truth this revelation had its origin over 30 years ago. The earliest teachings of the New Message, given in the 1980s, continually refer to the Greater Community, the larger panorama of life in the universe. The spirituality of life beyond our world was first revealed in a series of books which included *Greater Community Spirituality* whose 300 pages of mysterious revelation were given in just seven days. Years later, more was revealed in the first two sets of the *Allies of Humanity* Briefings, which speak of our destiny as a free world in the Greater Community. And then came *Life in the Universe.*

I remember the morning in 2008 when I was first called to receive it. It was early, before the sun had risen that I awoke suddenly, feeling a great Presence urging me to rise. I felt impelled to get up, leave home and walk half a mile to my personal study—a place I call the "Cloud Room." There I sat in the same chair where I had received many of the revelations of the New Message in years past. I sat quietly for a time. And then the Angelic voice filled my mind and the room with its great and overarching Presence and instructed me to begin. I started recording, unsure what would happen, having no personal sense of anything to say.

My mind was overcome. And then I faintly heard the first words emerge: "You live within a Greater Community of intelligent life in the universe. It is vast, encompassing the entire spectrum of evolution…"

On that day, I recorded the first five chapters of *Life in the Universe.* The next day I was woken again at sunrise and impelled to go to the Cloud Room and to continue. At that point, I knew I was in the process of receiving a book. That second morning, the next five chapters were spoken through me and recorded. And on the third morning, I woke very early. It was dark when I left home, walking on a footpath past the quiet, sleeping foothills of the Rocky Mountains.

That was the last morning. The final four chapters were given that day, ending with these words:

> "You are part of the Greater Community. You have always lived in the Greater Community. And now it is time for humanity to emerge into a Greater Community, which was always its destiny."

After each day and at the end of it all, I was completely exhausted. In just three days, without any mental preparation or preconception on my part, the book was completed. That was it. What you read in *Life in the Universe* are the exact words that were spoken in that brief span of time. This is a spoken revelation. The only liberty I have taken is to render the words in paragraphs and sentences so that they could be read more clearly and more easily translated into other languages. In your hands is the direct revelation from God, given through the Angelic Presence who oversee our world and all worlds in the Greater Community.

We have before us a doorway into our future and destiny in the universe. The universe has always been there—we have lived our entire lives immersed and surrounded by it. Yet most people have barely given a thought to what this might mean and how this greater panorama of life is now shaping our world and destiny.

Perhaps we look up at the stars to appreciate a particularly beautiful night sky. But then we quickly look back down; the fog of human concerns, human preoccupation and human beliefs is always waiting to re-envelop us. Yet the world is changing. Contact has begun, and our emergence into the Greater Community is underway. The doorway is open. Yet who will pass through?

Who can reconsider their own life and identity in light of this revelation? Who can consider the implications of contact with those forces who have arrived on our shores, promising peace, salvation and gifts from space, yet harboring hidden intentions. We are now the natives of a new world facing intervention from foreign powers. They are ready for us. Yet we are not ready for them.

There is so much to confuse and cloud our approach to the reality of life in the universe. Regarding Contact, we are immediately hopeful or fearful, immediately assuming who would come to our world and for what purpose. Despite our advances in science, our view into space is clouded by the lens of long-held beliefs, preferences and assumptions. In our isolation on Earth, we have had no reason or impetus to think about life in any other way. We project our values and perspectives out to space and from there assume what life could or should be like.

Life in the Universe teaches us that technology does not lead to higher ethics or morality. Though this truth has been clearly demonstrated in our own history, most people still prefer to entertain fantasies of enlightened space travelers arriving with pure altruism to offer salvation to a struggling humanity, or fears of predators and destroyers who come to wreak havoc with no limits. Both fantasies are far from reality.

Life in the Universe teaches us that as civilizations mature and become more technological, freedom often declines, if it ever existed to begin with. All advanced nations are in a constant and sometimes desperate search for resources. Technology does not free a nation from these resource requirements, but instead tends to increase them. There are limits to growth, everywhere and for everyone. Rather than free you, excessive technology chains you to a cycle of depletion and dependency that ultimately leads to environmental collapse and subjugation in the universe. This is revealed in *Life in the Universe.*

Much of this revelation runs against the grain of how we think about life. Yet what is the purpose of a revelation from God but to reveal, to correct and to educate? God is certainly not bound by our beliefs, preferences and assumptions, but only reveals what is truly needed for the welfare and advancement of humanity.

Only God can be the source of a revelation about life in the universe. Who else but God could speak about what religion means in its thousands and millions of expressions, in countless societies, at different ends of the universe? Not even the most advanced society in space could have direct knowledge of these things.

Without a revelation to prepare us for life in the universe, our freedom and sovereignty would be lost there. God knows this. That is why the New Message from God has been given. Humanity stands at the threshold of space. Our environment is in decline, our climate is becoming increasingly unstable, our nations and religions are fractured and the life-sustaining resources of Earth are being rapidly depleted. Intervention from forces in the universe is underway. Can you see what is at stake? Can you see our great need for God to speak again?

The New Message from God, of which *Life in the Universe* is a part, is an answer to the great needs prevailing in the world today. Part of its purpose is to protect humanity, to prepare us for Contact and to spark the awareness that we are one people with one destiny, stewards of a beautiful planet that is greatly valued by others beyond our shores.

What else could finally bring an end to our ceaseless conflicts and pave the way for greater freedom, cooperation and unity on Earth? We cannot be fighting each other if we are facing greater danger and adversity from the universe beyond. We cannot afford endless conflict over who will have access to the remaining resources of the world. We have great cause to unite for our own mutual survival, security and well-being.

Another purpose of God's New Message is to reveal the deeper mind that God has imbued in all sentient beings, a deeper mind called Knowledge. Knowledge represents the core of our spirituality, yet its presence is remarkably absent in many of the world's religious teachings. Through Knowledge, God is calling each of us out of the darkness of our past into a new life of contribution, relationship and self-realization.

In the universe, religion is represented by Greater Community Spirituality, an experiential and elemental spiritual force and reality that can be translated from one world to another free of culture, creation stories, hero worship or the particular history of any tribe or group. For the first time in history, the spirituality of the universe is now available here on Earth.

Some will question if we are really worth it. Living on Earth, itself but a speck in a universe impossibly huge, humanity continues to be mired in conflict, degradation, violence and division. Are we deserving of a new revelation, a new beginning and a new way forward?

Yes, we are. Despite our many errors and the tragedy of much of our history, humanity has kept Knowledge and spirituality alive. In a universe of civilizations where these have so often been forgotten and denied, this a great achievement. The power of Spirit and individual freedom have not been lost here, not yet.

The impact of contact with the Greater Community is complete. It can either completely overtake us, or it can completely renew, strengthen and unite us as never before. Which road will we take: the path of submission or the path of freedom and self-determination?

Humanity has great promise. We are worthy of this new future, a future that will be unlike the past. Let us deliver upon the promise of humanity. Let us receive this revelation, realizing our great need and the great answer that has come.

Let us build and establish our place in the Greater Community as a free and self-determined people and resist the influences being placed upon us by those few races who seek to use the world for their own purposes. Let us become allies of humanity and safeguard human civilization from exploitation and corruption from within and intervention from without. This is a human sovereignty movement on Earth, and the New Message is our mission statement, our roadmap and our guide into the unknown.

This is the time to deliver upon the promise of humanity. We are no longer an isolated tribe in an unknown world in the universe. Our isolation is over. We stand facing a Greater Community of worlds, a non-human universe where individual freedom is rare. This is the greatest event in human history and for this we must prepare.

If you feel called to a higher purpose, and if you feel connected to life beyond the world, then this book is for you. *Life in the Universe* is for you. Step through this doorway with me into a greater universe and a greater life.

Marshall Vian Summers
2012

Join The Preparation

Join a worldwide community of people who are pioneering a new chapter in the human experience, one where we protect and preserve our world and prepare for the greatest event in human history: our emergence into a Greater Community of intelligent life in the universe.

Learn more about the free educational opportunities available and meet the thousands of people worldwide who are preparing for life in the universe.

-Forum-

-Free School-

-Worldwide Broadcasts and International Events-

-Annual Encampment-

-Online Library and Study Pathway-

www.newmessage.org

Printed in the USA
CPSIA information can be obtained
at www.ICGtesting.com
LVHW041924260124
769459LV00002B/2